# HELICOPTER MATRIX

# Patrick Flannery

# In loving memory

To my mother

A woman who lived by faith and

not by sight

Flying a veering
helicopter
Is much like being a
Dragonfly


Patrick Flannery

# What's in the book

# Before anything else

Bird's bodies are made to fly. They have Light bones and specially shaped wings. The act of flying helps the birds get away from animals. It makes them better hunters. Flying takes them traveling from places to places. That's the science behind things.

 The birds are using their legs to push off the ground into the air. We call that - thrust.

Birds must flap their wings to stay flying. We call that - a lift.

Their wings have a bowed shape, like a spoon. This wing's shape helps them in flight as the air moves above and below their wings and keeps them in the air. As air rushes through their feathers, it helps to create more lift. When you add thrust and lift, the birds are flying.

Birds don't flap their wings at all times when they fly. When they're up there, high in the sky, they could also glide, which means getting a free ride in the air. All they need to do is flaps their wings once a while.

They can also let an air jet push them up higher, to the tops of really tall trees. We call it - climbing. Birds fly in many different ways. A sea bird can dive fast to catch a fish. Hummingbirds flap their wings so fast all you can see is a blurry motion.

The free acrobatic of the dragonfly kinematic flight is amazing to witness. They are one of nature's most perfectly engineered creatures. They can fly with ease, backward, forewords left and right. It's described as a phenomenon.

They can perform a clap and fly, in the proximity to which the wings emerged each other during this maneuver with the total force produced during their wing stroke. The dragonfly beat its wings with a set of stroke planes respective to the length of their body-axis; they aligned their stroke planes nearly too normal the thrust force's direction. To achieve this, the dragonfly body alignment is correlated with the direction of the thrust.

You have to thank the birds for showing humans how to fly.

# HELICOPTERS

A helicopter is in a category of aircraft that utilized rotating wings to fly. Unlike an airplane or glider, the helicopter wings are spinning. Since the helicopter is heavier than air, it must use an engine to operate. The helicopter's spinning rotor allows it to perform tasks an airplane cannot. The rotor blades create lift as they spin. An airplane must fly fast to move enough air over its wings to provide lift (moving the air above it).

The helicopter may take-off or land without the need for any landing runway. With no resemblance to an airplane, a helicopter can fly backward or sideways, hover at one point in the air without moving forward. It makes helicopters perfect for accomplishing missions an aircraft unable to do.

Rules regarding helicopter landings in the US are much more relaxed than those for winged aircraft.

No law says that helicopters need helipads to land.

The FAA must be notified regarding all permanent landing areas.

Helipads must meet the design requirements of the FAA.

Thanks to the helicopter's ability to take off and land characteristics vertically, and its ability to hover for extended periods, it became the best choice for an assignment that is not possible otherwise.

The uses of helicopters include shipping cargo, armed forces, fire, medical, and aerial search observation.

Some helicopters can fly up to 30,000 feet, but they are deliberately designed to have this additional capability. The same helicopter can hover at 15,000 to 18,000 feet because they do not have the added lift from a forward speed. The low speed is the main limitation of helicopters.

A helicopter cannot fly as fast, and there several reasons for that.

The complex aerodynamic airflow through the wings is still the most difficult phenomenon to explain.

The main principle of the vertical flight is the airflow through the rotating wings. Rotating-wing aircraft are very complicated and sophisticated in their operations. Apart from the airflow, is the instability and coupling of the helicopter system.

Helicopters come in many sizes and shapes, but most of them share the same significant apparatuses. These apparatuses comprise a cabin where the payload and crew are carried. The engine's power transfers directly to the transmission, and then into the main rotor and the tail rotor. The rotor system found on helicopters contains a single main rotor or dual rotor. Most of the dual rotors turn opposite each other, so the rotating force of the rotor opposes the rotating force from the other rotor, and the spinning inclination will cancel.

The rotor system usually consists of two or more rotor blades. The blades can flap, feather, and lead or lag separately of each other. Each of the rotor blades is attached to the rotor hub by a horizontal hinge, known as the flapping hinge, which permits the blades to flap up and down separately.

The flapping hinge is located at varying distances from the hub, and there may be more than one. Each manufacturer chooses a different position. Each rotor blade is attached to the hub by a vertical hinge, known as a drag force or the lag hinge that permits each blade to move back and forth.

The Dampers are usually integrated with the rotor system's design to prevent too much motion about the drag force hinge.

The drag force hinges and dampers' purpose is to absorb the rotor blades' acceleration and deceleration. Feathering means changing the pitch angle of the rotor blades. By changing the blades' pitch angle, you can control the main rotor disc's thrust and direction.

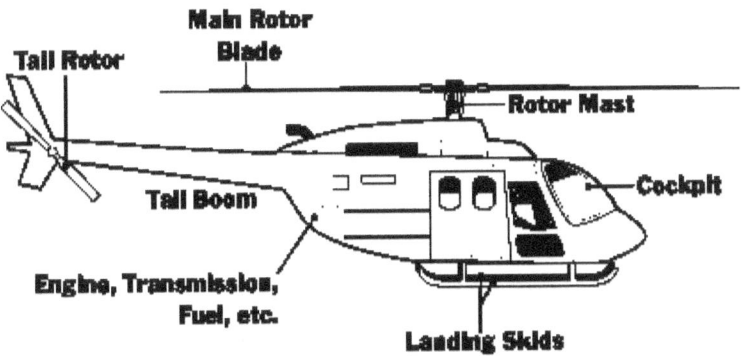

As the engine turns in a single main rotor helicopter, it creates a rotating force that causes the helicopter's body to turn in the opposite direction of the rotor.

To counter the Ant-rotating force, you must allow the helicopter to maintain its heading and prevent it from wobbling. It is possible to make an emergency landing in a helicopter should a tail rotor failure occur.

The procedure is the same as that of an engine failure, and it's called an autorotative landing. Essentially a powerless helicopter becomes an auto-gyro (gyrocopter) in flight and can be flown as such.

# The Robinson R-22

The R-22 helicopter is a central cause of the idea of excellent design. In the R-22 class, no other helicopter march or equals in reliability, speed, or low operating cost.

The two-seater R-22 has excellent performance, exceptional reliability, and ideal for flight training.

It has two-blades, single-engine, and considers a light utility helicopter manufactured. More than 8000 R-22 helicopters are used internationally. In its class, the helicopter holds significant performance records, which includes speed and distances.

The Beta II model's clean aerodynamic design allows a top speed of 100mph or more and burns less than nine gallons of fuel per hour.

The cabin provides comfortable seating for two with removable dual controls on the passenger side. Instead of floor-mounted cyclic sticks between the helicopter pilot's knees, the R22 uses a unique T-Bar control connected to a rod that emerges from the console between the seats.

Push-to-talk switch intercom communications activated for the left seat occupant by a floor-mounted foot.

Some later models may also equip with a voice-activated intercom system.

The helicopter rotor system consists of a two-bladed main rotor and a two-bladed anti-torque rotor on the tail: the Collective and cyclic pitch inputs are connected through push rods conventional swashplate mechanism.

As a pilot, you need to make small adjustments by twisting the throttle on the collective throughout the flight. The R22 uses a four-cylinder, air-cooled, normally aspirated engine.

The two-blade rotor in the small size R22 makes it possible to transport the helicopter without blade folding.

# The Robinson R-44

The Robinson R44 is a four-seater light helicopter. Its features assisted hydraulic flight controls, and probably the best helicopter for private possession and training.

The R44's steel-tube structure provides a rigid yet lightweight airframe. Due to the R44's open cabin design, the aerodynamic fuselage optimizes the airspeed and fuel economy.

Four passengers can sit comfortably in the R44, without a view obstruction. A low tail-rotor tip speed and large tail decreasing the fly-over noises, resulting in a community-friendly helicopter.

# Bell helicopter

The Bell 204S and 205S light utility helicopters converted to civilian versions from the UH-1H that is used in the military. The helicopter has a metal fuselage construction with tubular landing skids and two rotor blades on the main rotor. Several models featured a single turboshaft engine, and the later models feature twin engines with four-blade rotors. All helicopters in the bell family have similar construction. The main structure comprises two longitudinal main beams under the passenger cabin forwarded to the nose and back to the tail boom. These main beams, separated by partitions, support the cabin, landing gear, the engine, and the tail boom.

The Seats in the cabin arrange with two helicopter pilot seats in the front and additional seating in the back to 13 passengers or aircrew. The passenger seats were formed with aluminum tubes and canvas material and could be removed quickly.

The access to the cabin is through the two sliding doors. The doors could be removed. The collective lever, anti-torque pedals, and the cyclic stick comprise one hydraulic system with dual controls.

As the Air Force dubbed it, the new Combat Rescue Helicopter comes with a suite of improvements over the HH-60 Pave Hawk it's replacing, and some of them long needed. It will able to fly farther and protect itself better.

It contains new mission-planning hardware that allows crews to coordinate their efforts better and more efficiently search for injured comrades.

Its mission-planning hardware provides multifunction displays not just for the pilots but for the crews in back, who help coordinate and execute rescue missions. Often, the flight engineer in the back wants to know the altitude, airspeed, direction, and other data or look at the infrared displays, but requesting that information from the pilots will take precious seconds away from their ability to safely fly.

In the rear, the crew also gets newly configured seats that swivel to face either those multifunction displays, inside toward patients on stretchers, or outside to fire the window-mounted machine guns. The seats can also collapse and stowed up at the ceiling in seconds, to create more room.

Stretchers can stack and configured to carry more or otherwise optimize their position for the rescue crews working on patients. A crew would include two pilots up front, a flight engineer and a paramedic in back, and room for several litters.

# Sikorsky HH-60W Rescue Helicopter

The HH-60W has roughly twice the fuel capacity as its predecessor, thanks to a new tank positioned behind the cabin that replaces the Pave Hawk's bulky auxiliary tanks, which took up interior space. That wills double the flight time from about two and a half hours to five, without impacting the helicopter's balance. The extra range means it can fly into combat zones the Pave Hawks can't reach, an important factor in light of how combat potentially shifting from the guerrilla tactics of Afghanistan and Iraq to more modernized forces in Iran and Syria.

Compared to many warbirds, the US Air Force's newest helicopter seems positively peaceful. It carries no Hydra rockets or Hellfire missiles, just a pair of .50-caliber machine guns—you can't float about totally unarmed, after all. That's because this chopper prioritizes saving lives overtaking them. Sikorsky's HH-60W is a flying ambulance, a medical evacuation aircraft designed to dip into combat zones, pile up the injured, and whisk them to safety.

Instead of using precious lifting capability for missiles and rockets, it hauls extra fuel, configurable stretchers, and a hoist to help wounded men out of areas where the chopper can't land.

The new Combat Rescue Helicopter comes with a suite of improvements over the HH-60 Pave Hawk it's replacing, some of them long needed. It will able to fly farther and protect itself better.

It contains new mission-planning hardware that allows crews to coordinate their efforts better and more efficiently search for injured comrades. Its mission-planning equipment provides multifunction displays not just for the pilots but for the crews in back, who help coordinate and execute rescue missions.

The seats can also collapse and stowed up at the ceiling in seconds, to create more room. Stretchers can stack and configured to carry more or otherwise optimize their position for the rescue crews working on the patients. A pivotal crew would include two pilots up front, a flight engineer and a paramedic in back, and room for several litters.

The latest helicopters also come equipped with a Tactical Mission Kit, which includes advanced detection and defensive systems; hardening against cyberattack, improved connectivity with other networks, vehicles, and troops in the area; new optical and infrared cameras; and missile- and rocket-detection capabilities.

The new HH-60Ws Equip with all the proper sensors. Supporting this approach to the HH-60W, there are also doing a rapid upgrade to the HH-60G.

# Sikorsky S-76

The Sikorsky S-76 model helicopters medium-size utility helicopter is virtuously designed for a commercial rather than military use.

The S-76 comes with a few different models, but they all have the same outline, with a fully articulated rotor and a four-bladed anti-torque tail-rotor on the boom. The helicopter comes with two turboshaft engines positioned above the passenger cabin.

The fuselage made of metal composite, a fiberglass nose, and a light alloy honeycomb cabin — the tail boom made of light alloy.

Two helicopter pilots (or a helicopter pilot and a passenger) sit side-by-side in the cockpit, ahead of the main cabin, which provides accommodations up to 12 passengers in three rows of four seats. In luxurious executive helicopters, you could find four to eight rows.

It has the new version of the Rolls Royse engine, the Allison 250, which was developed especially for the S-76, with a single centrifugal compressor, instead of the multi-stage centrifugal compressor of earlier models.

Other models, the S-76 series, incorporate by Pratt & Whitney's new engine. The helicopter has an enhanced ground proximity warning system (EGPWS), navigation management system, weather radar, and optional rotor anti-ice system.

The main rotor hub has a single-piece aluminum hub with elastomeric bearings, which does not require lubrication or maintenance. The main rotor blades have titanium spars with a 10-degree twist to allow even loading when hovering. The rotors tips tapered and swept back.

The S-76A has a retractable nose wheel to increase its cruising speed and is equipped with flotation bags filled with helium in a forced emergency landing on water.

More than 800 S-76 helicopters have been serving multi-mission' role, which comprises offshore oil, VIP transportation services of a head of state, emergency medical services, Search and Rescue, and other civil defenses.

# MD 520n NOTAR

The MD 520N helicopter has a fully articulated five-bladed rotor system that provides excellent control and maneuverability characteristics. With its advanced NOTAR anti-torque system, the helicopter is a member of an exclusive class of the safest, quietest helicopters worldwide.

The MD 520N is one of three helicopter models, factory-made by MD Helicopters. The exclusive NOTAR system. By replacing conventional tail rotors with the NOTAR system, the margin of safety increased for the crew. It reduced noise level, making it a good choice for operation in populated zones.

The MD 520N is fast, lightweight, and a turbine-powered helicopter. It uses advanced technology design of the helicopter, resulting in excellent speed abilities and handling performance capabilities.

For military helicopters, the lower vulnerability and lower radar return would seem like big advantages.

# The speed of a helicopter

A drag force is opposing the thrust force. In helicopters, the parasitic drag force formed and presents any time the helicopter moving through the air. This Drag force is opposing thrust by reducing the amount of thrust available to fly helicopter faster. It includes the landing gears, doors, and the fuselage, which shapes and produces parasite drag force.

On the late model helicopters, the landing gear is retractable to reduce parasite drag force production. When the helicopter moves forward, the airflow over the advancing rotor' blade increases by the forward speed as the blades move in the opposite direction of the flight.

A region near the blade roots on the retreating side experiences a reversed flow as the helicopter's forward speed increases. Due to these limiting factors, the maximum speed of helicopters is limited to 250 mph. An average helicopter manages to break 110mph. A Blackhawk can do about 120mph. Others cannot even reach 100. Fast helicopters barely break 200mph. The current helicopter high-speed record was held by the Westland Lynx at 402 km/h (250mph) using specially designed high-speed rotor blades, which the maximum speed of a helicopter.

The X3 helicopter bit the previous record for helicopters by cruising at a speed of 291 **MPH**.

Even though the retreating blade limits a helicopter's forward speed, their angle of attack shapes the limiting factor and not the rotor blade airspeed.

Moreover, because the blade stall occurs ninety degrees to the longitudinal axis, the rotor's downward force will cause a pitch up but not a roll.

Typically, an engine supplies the power to the main rotor through transmission and a belt drive or centrifugal clutch system. A small helicopter has a reciprocating engine, which secures to the airframe. The engine could be situated horizontally or vertically.

# HELICOPTER STRUCTURE

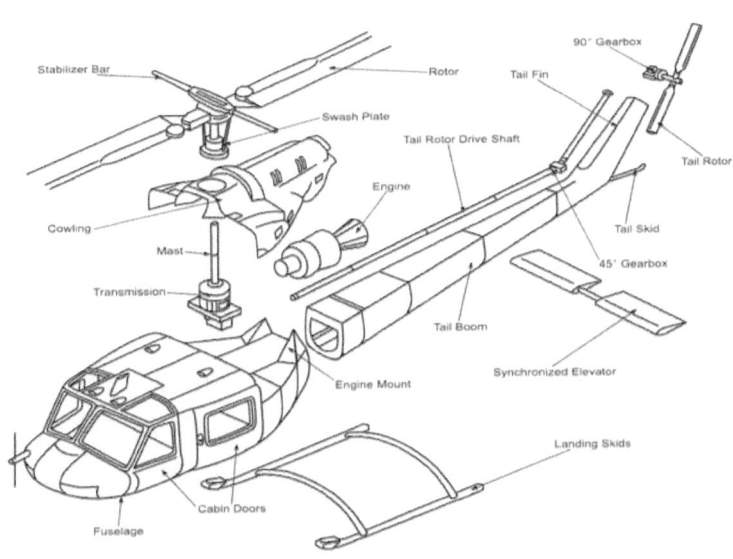

The airframe is the fundamental structure of a helicopter, made of either metal or other material, and has several purposes. The airframe's primary task is to hold all the components in their respective positions and allow the engine to transfer the power to the main rotor, the tail rotor, and the landing gear.

The airframe's outer core, the fuselage, contains an aircraft's body section that accommodates the cabin, which holds the crew, passengers, and cargo.

All fuselage structures are based on the same principle; they aerodynamics and titanium in areas that are subject to higher stress or excessive heat. Seating arrangements in a helicopter cabin vary. The fuselage also accommodates the avionics, the flight controls, and the power plant.

The cabin for the crew is made of fiberglass or composite plastic materials. The frame's design depends on the helicopter's model, but a tube structure is prevailing in most cases. The helicopter's tail is a stressed-skin structure fixed on a series of alloy tubes or steel rods.

Most helicopters have one collective for each seat, located on the left, so the pilot wants to use the hand in the middle of the cockpit to work the radios and things. Hence, they sit on the right side, which places the collective hand in the middle of the cockpit.

Helicopters could have floating devices for water maneuvers or skis for landing on snow or other soft terrains. The helicopter's wheels are also a type of landing gears, either tricycle or a four-point configuration. Usually, the nose and the tail gear is free to swivel when the helicopter taxiing on the ground.

# The main rotors

A helicopter's main rotor system is acting as an aircraft's wings. Because the rotors airfoils, the lift acts as much as an aircraft's wing and reacts to changes in the angle of attack, or stall, just like wings.

In lift in an aircraft, it is generated by the wings, and the thrust generated by its propeller. Lift and thrust in a helicopter are generated by the same exact component: the main rotor blades. The rotor thrusts the air downward, and the helicopter ascending. If the rotor is tilted, the helicopter will move in the direction of the tilt. As the rotor turn, the opposite force will move the helicopter's fuselage in the other direction. A separate rotor is required to overcome the rotating force in most helicopters with a single main rotor. It's the tail rotor which is compensating this twisting force.

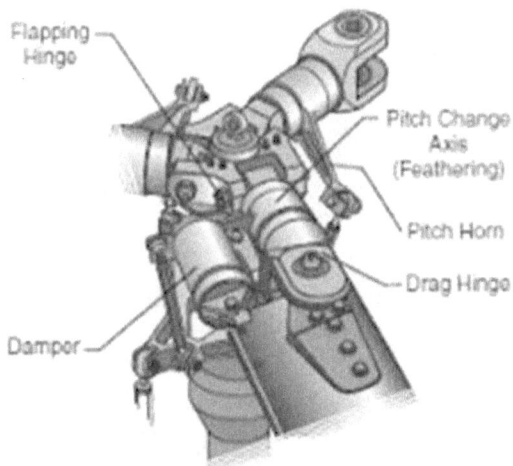

# Main Rotor System

The main rotor system is the part of a helicopter that generates a lift in the main rotor. The main rotors provide a vertical lift, and it may fix horizontally in the helicopter. A tail rotor is fixed vertically to provide horizontal lift as thrust, counteracting the main rotors' rotating force effect.

The main rotor entails a mast, hub, and the rotors. The mast is constructed as a hollow cylindrical metal shaft, which extends upwards and a cylindrical metal shaft, which extends upward. The attachment points for the rotor blades the hub at the top of the mast. The rotor blades are attached to the hub. It's an airfoil with a very high aspect ratio. The blades' pitch is controlled by a swashplate connected to the helicopter's controls — the design of the rotor aims to operate in a very narrow range of speed.

The main rotors hinged to the rotor head to have limited movement up or down, and they can change their pitch (angle of incidence).

The blade on a helicopter has a large diameter that allows it to capture a large air volume. The large volumes of air allow a lower downwash speed and thrust. It's more efficient at low speeds when accelerating a large amount of air than a small amount of air.

The helicopter's main rotor may have up to six blades, depending on the design—the controls for the main rotor known as Collective and Cyclic Controls.

Three basic classifications of rotors in the helicopters are the semi-rigid, rigid, or fully articulated system. Some rotors use a combination of these types.

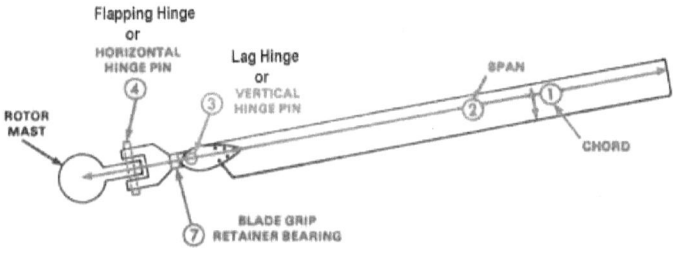

**Main Rotor Blade System**

The main helicopter rotor or the rotor system is a combination of a rotary-wing and a control system that generates the aerodynamic lift force.

A semi-rigid rotor system has two main rotor blades. The blades flaps move up and down as one unit, and they can feather-twist at the rotor hub as one unit to adjust the angle of attack as they are rotating.

Single-rotor helicopters entail a tail rotor to counteract the twisting momentum of the single large main rotor. Tandem rotor helicopters use two counter-rotating main rotors, with each of them canceling out the other's rotating force. Counter-rotating rotors won't collide or destroy each other if they flex into each other pathway.

Tandem rotor helicopters, occasionally referred to as dual-rotor, and they have two large horizontal rotor assemblies.

This configuration has the advantage of being able to hold more weight and shorter blades, as there are two sets. Besides, all of the engines' power can be used for lift, whereas a single rotor helicopter uses it to counter the rotating force.

Thus, the tandem helicopters are considered to be among the most powerful and fastest in the world.

A set of two rotors that are turning in opposite directions is called an Intermeshing rotor. Those are two rotors turning in opposite directions but fixed together on one mast and have the same rotation axis one above the other.

The Mast is held at a slight angle to prevent the rotor-blades from colliding and correctly intermesh. This arrangement allows the helicopter to function without the need for a tail rotor.

# Semi-rigid Rotor System

A semi-rigid rotor system allows two different kinds of movements, feathering and flapping. The semi-rigid system contained two blades attached to the rotor hub. For the main rotor, the hub is free to tilt on the teetering hinge. The teetering hinge permits the main rotor hub to tilt, and the feathering hinge enables the pitch angle to change.

The underslung rotor system mitigates the lead/lag forces, so the lead and lag forces are minimized. It's accomplished by mounting the rotor blades slightly lower than the usual rotation. As one of the blade flaps down, the other blade will flap up. A Feathering occurs when the feathering hinge changes the pitch angle of the blade.

The semi-rigid system's flapping characteristics can lead to the main rotor hub touching the main rotor mast. It may a result of too much rotor flapping.

The main rotors are designed to achieve maximum flapping angle. Under usual flight, the condition of blade flapping is minimal, in straight and at level flight, perhaps 2°. With high straight-forward speeds, the flapping angles will increase slightly at the low rotor speed, at high gross weights, and turbulence. Maneuvering the aircraft during low-speed flight at extreme CG positions can induce larger flapping angles.

# Rigid Rotor System

The Rigid rotor systems are expensive to produce. Typically, it is constructed out of titanium and other composites material. The systems require less maintenance than other rotor systems but resist the phase of the aerodynamic effect of the low-G condition.

The term "rigid rotor" usually refers to a hinge, less rotor system with blades flexibly attached to the hub. Loads from flapping and lead/lag forces are accommodated through rotor blades flexing, rather than through hinges.

The rigid rotor system is mechanically simple but structurally complex because of the operating loads the system must absorb by bending rather than by hinges. Here, the blade roots are rigid-fixed to the rotor's hub. Rigid rotor systems tend to behave like a fully articulated system due to its aerodynamics.

As helicopter aerodynamics and materials engineering continue to improve, the rigid rotor systems may become more common as the system inherently easier to design and offers the best properties of semi-rigid and fully articulated systems.

The rigid rotor system is very reactive, and it's not likely to strike the mast like semi-rigid systems could because the rotor hubs are securely attached to the main rotor mast.

It allows the rotor and the fuselage to move together as a single entity and eliminates much of the oscillation usually present in other rotor systems.

Since there no hinges to assist and absorb the larger loads, vibrations could feel throughout the cabin much more than on other rotor head designs. There several variations of the basic three-rotor head design.

# Fully articulated system

In the fully articulated system, each of the rotor blades gets attached to the rotor hub through a series of hinges that let the blade move independently. The system has more than three blades. The blades can flap or feather independently.

Horizontal hinges permit the blade to move up and down, and the flapping hinges may be located at a distance from the rotor hub. The Vertical hinge permits the blade to move back and forth. Dampers are usually used to prevent excess movement around the drag force hinge. The purpose of the drag force hinges and dampers to compensate for acceleration and deceleration caused by the Coriolis Effect. The lift force's magnitude is based on the Collective input, which changes the pitch on all blades, the same direction and at the same time.

This system more expensive to manufacture and maintain than semi-rigid rotors but less susceptible to low-G conditions and bumps. However, it is greatly affected by ground resonance.

# Rotor Blade twist

Because of the potential lift differential along the blade resulting primarily from speed variation, the blades are designed with a twist. Blade twist provides a higher pitch angle at the root where the speed is low, and a lower pitch angles nearer the tip where speed is higher. This design helps to distribute the lift more evenly along the blade. It increases both the induced air velocity and the blade loading near the inboard section of the blade.

Two angles are enabling a rotor system to produce the lift required for a helicopter to fly. A helicopter in a straight forward motion or hovering with a headwind or crosswind, has more air molecules entering the aft portion of the rotor blade. At that time, the attack angle is lesser, as the induced flow is greater at the rotor disk's rear.

The relative wind moves in parallel but in opposite direction to the airfoil's movement, striking the blade at 90° to the leading edge and parallel to the rotation plane. It constantly changes direction during the rotation.

The rotational relative wind-velocity is the highest at the blade tips, and it decreasing uniformly all the way to zero at the axis of rotation (center of the mast).

At the onset of a blade stall vibration, you should reduce the collective pitch and increase the rotor speed to reduce the forward airspeed to minimize maneuvering.

The major warnings of a retreating blade stalling in low-frequency vibrations are equal to the number of blades per revolution of the main rotor system, pitch up of the nose, and the helicopter's tendency to roll towards the stalled (retreating blade) side of the rotor system.

When a three-bladed rotor system flaps upward, the center mass of that blade moves closer to the rotation axis, and the blade will go faster. When the blade flaps downward, its center of mass moves further from the axis of rotation, and the blade slows. This will increase and decrease the blade rotation due to the Coriolis Effect.

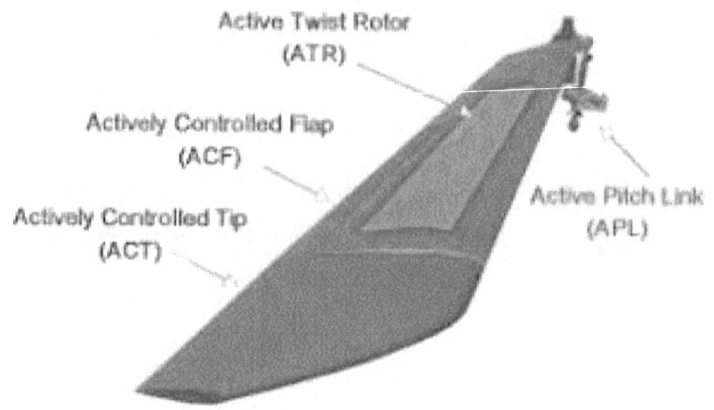

Active Twist Rotor
(ATR)

Actively Controlled Flap
(ACF)

Actively Controlled Tip
(ACT)

Active Pitch Link
(APL)

# Rotor Blade rotation

The rotor blade's base part is referred to as the stall region, and it operates above its maximum AOA stall angle. It causes a drag force, which tends to slow the rotation of the blade. With lower AOA on the blade's advancing side, most of the blade will fall in the driven region. On the retreating side, most of the blade is in the stall part.

A small segment near the root has a reversed flow. Thus, the size of the driven region on the retreating side will be reduced. Before landing from an autorotative descent or autorotation, you must flare the helicopter to decelerate. You need to initiate the flare by applying aft to the Cyclic.

If you fly a helicopter with a clockwise rotation, the left and right references must be reversed.

When you transition forward from a hover or take-off from the ground, you must be aware that as the helicopter speed increases, the translational lift becomes more effective and causes the nose to rise or pitch up (blowback). You must correct this tendency to maintain a constant rotor disc altitude, which will make the helicopter advance in speed range where the backlash occurs.

If the nose is allowed to pitch up through this speed range, the aircraft may tend to roll to the right. The correct action is to continuously advance the Cyclic as the helicopter's speed increases until the takeoff is completed, and the helicopter has switched to a transfer flight.

In hover, the cyclic lever is centered, and the pitch angle on the blades, which move forwards and backward, remains the same.

Moving the Cyclic forward at low speed will reduce the pitch angle on the advancing blade and increase the backward blade's pitch angle.

This will cause the rotor to tilt slightly. At a higher forward speed, you must continue to advance the Cyclic. As an outcome, there will be more tilt relative to the rotor than at a lower speed. A horizontal lift component (thrust) is generating a higher airspeed for the helicopter. The higher speed induces the beat of the blade to maintain the symmetry of the lift.

The combination of a flutter and a cyclic feathering maintains the lift's symmetry and desired altitude on the rotor system and the helicopter.

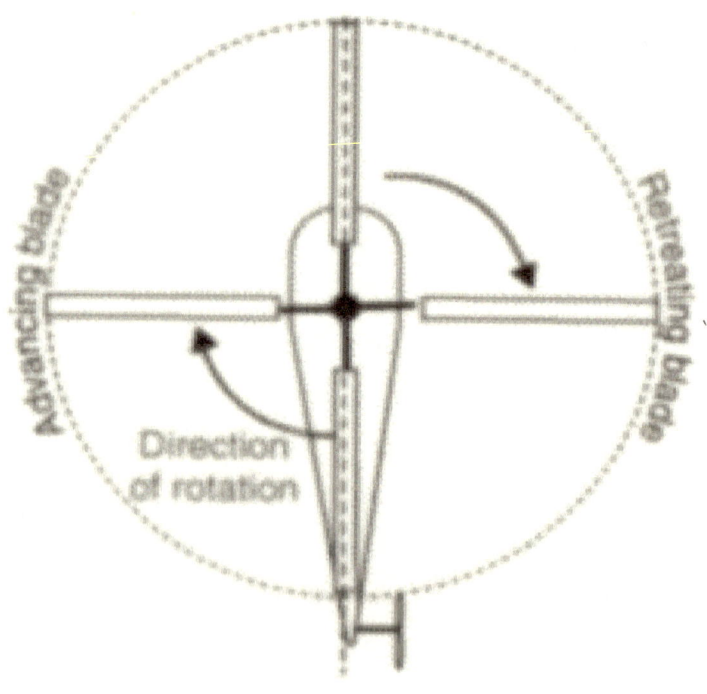

# Swash Plate

A Swashplate contains two main parts: the stationary swashplate and the rotating swashplate. A series of pushrods controls by the cyclic and the collective are connected to the stationary swashplate secured around the main rotor mast. Both swash plates can tilt, slide up or down as one unit. The swashplate aims to translate the inputs from you as a pilot into the rotor blades' rotating inputs. The swashplate can tilt in all directions and could move vertically, but it's restrained from rotating by an ant drive link. The rotating swashplate is fixed to the stationary swashplate by a sleeve, connected to the mast by drive links, and rotating with the main rotor mast.

1 Non-rotating outer ring

2 Turning inner ring

3 Ball joints

4 Control (pitch) preventing turning of an outer ring

5 Control (roll)

6 Linkage to the rotor blade

The pitch links or universal joints connect the rotating swashplate. If the engine fails, these airfoils must free to rotate to provide the lift to the helicopter. The freewheeling unit is automatically disengaging the engine from the main rotor when the engine revolutions per minute (RPM) less than the main rotor RPM. The most common freewheeling unit assembly comprises a one-way clutch located between the engine and main transmission.

# Landing Gears - skids/wheel

When it comes to landing gear, helicopters have two basic types: skids and wheels. Skid gears are always fixed, and wheels can be fixed or retractable. Larger helicopters (such as the Bell 206 or AS350) can be moved around with ground-handling wheels, but it takes a couple of people to move it.

On more powerful twin-engine and larger helicopters, weight may not be such a concern, and then retractable wheels can make more sense. One of the benefits to wheels is its ability to taxi conveniently.

The decision to use fixed skids or wheels is based on how the helicopter will be used. There are a wide variety of choices. For example, the Bell 429 helicopter come with wheels, but can also come with skids.

As with many of the configurations to be made when selecting the right helicopter, defining the missions, your helicopter will be used for will most likely be the defining factor.

Landing gear usually includes wheels equipped with simple shock absorbers, or more advanced air/oil struts, for runway and rough terrain **landing**.

The skid type landing gears are the most common, and it's suitable for landing on multiple surfaces. Skid gears are armed with shocks to observe the touchdown's jolts from spreading to the fuselage system. Other skid type's gears absorb the jolts by bending the skid attachment arms. The landing skids are fitted with replaceable heavy-duty shoes to protect them from wear and tear.

Skids weigh less and are mainly used because they are cheaper and lighter than wheels. Use wheels because a skid-equipped helicopter cannot move on the ground and need a set of ground-handling wheels to jack up and rolled away(for maintenance). Also, when the pilot shut down the engine, the helicopter could wander-off. Therefore, it's more convenient to move the helicopter with permanent wheels. The larger helicopters must have them.

The design of the retractable wheels is complex, but the helicopter could enjoy increased aerodynamic efficiency. The fixed gears are simply designed, but causes increased drag force and reduced efficiency.

A skid landing gear needs very little maintenance.

If the helicopter's primary mission is medical or transportation, the folding wheels will allow a greater speed over long distances and increase the fuel economy.

# THE POWERTRAIN

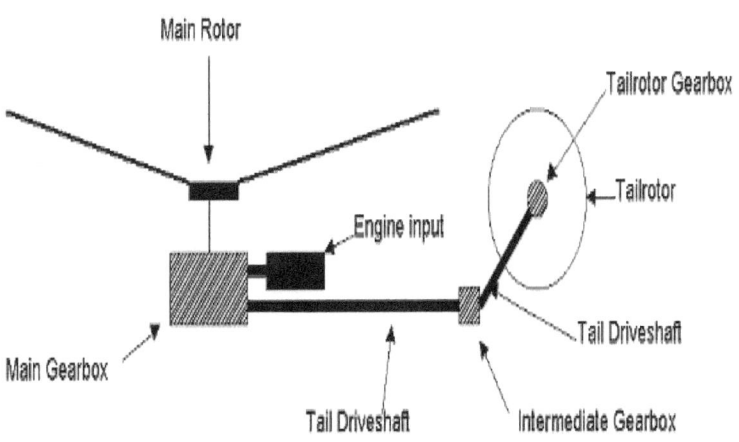

# The Engine

Jet engines are used as a driving power and thrust production. Turboshaft or turboprop engines, or jet engines in helicopters, do not produce thrust. They simply turn energy into a mechanical power that drives motors or propellers.

The turbine blades shaped are determined whether they change thrust into power to drive a shaft or allow the thrust for propulsion while taking a small amount of the drive power away to drive the engine's compressor section.

Some helicopters use the turboshaft jet engines to power the rotors. The Huey, Jet Ranger, and the Apache use the gas-turbine engines. Some of the older or smaller helicopters use 'reciprocating' (Piston) engines for their power source, but most of the helicopters in use today use gas-turbine engines.

Those engines are light, very powerful, cost-effective, and reliable. The gas-turbine engine failure rate is very low because many internal moving parts are reciprocating.

A jet engine's working principles are suction (intake), compression, Burn, and Exhaust.

# The compressor section

The engine draws-in air and compresses it to make it denser and better for combustion. By turning the compressor blades, which are shaped like small airfoils, air enters into the compressor. The air drowning between the moving rows of the compressor blades through stationary blade sets we call - Stators.

The stators help in the compression process by changing the airflow direction. The air starts occupying less volume as it moves through the compressor.

The air then moves through a diffuser section, which transports the air into the combustion chambers in the combustion section. Then the air is mixed with fuel and ignited to create the 'Burning' portion.

The burned fuel then travels into the turbine section, where the force is turning into a drive power and a thrust.

In most helicopters, the power of a turbo-shaft engine drives the transmission. If the engine's power only drives the compressor and directs the rest for creating thrust, it is considered a turbo-jet.

These engines are used in most medium to heavy-lift helicopters due to their large horsepower output.

# Transmission

From the engine, the power is transferring through the gearbox and then to the transmission. In the transmission, the engine speed reduced from thousands of **RPM** engine speed to only hundreds of **RPM**. Therefore, the torque increases as the rotor slow down to an acceptable level for that rotor system.

The transmission drives the mast, which provides a direct rotation to the rotors. Often enough, there another shaft out of the transmission for driving the tail rotor.

The accessory gearbox, which is mounted on the engine, draws little engine power to drive the oil pump, alternator, and the fuel control of the engine itself.

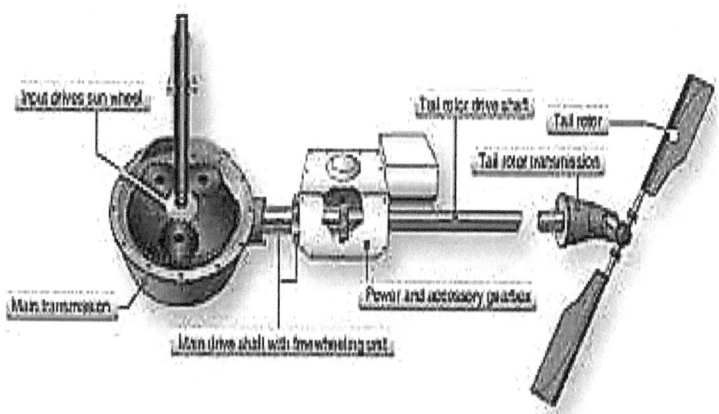

# Centrifugal Clutch

The centrifugal clutch has a built-in inner assembly and an outer drum. The inner assembly connects via the driveshaft and contains linings, similar to the automotive brake shoe material. The springs hold the shoes in at low engine speeds, so there is no contact with the outer drum attached to the transmission input shaft. As the engine speed increases, the centrifugal force causes the clutch shoes to move outward and slide against the outer drum. At this point, the transmission input shaft begins to rotate, causing the rotor to turn slowly. The rotor increases its speed as the friction increases between the clutch shoes and the transmission drum.

The rotor tachometer needle shows the increase by moving toward the engine tachometer needle. The engine engages the clutch through centrifugal force. The driveshaft turns at the same rate as the rotor unless a rotor brake is used to separate the main driveshaft's automatic engagement. By that, there is a negligible effect on the CG as the fuel is burning- off.

Most helicopters use hydraulic actuators incorporate in the system to overcome the higher control forces. The hydraulic system consists of actuators or servos and a pump, usually driven by the transmission, and a reservoir to store the hydraulic fluid.

Some helicopters have accumulators on the pressure side of the hydraulic system. It allows continuous fluid pressure in the system.

Inside the cockpit, a mounted switch is installed to turn the system off as needed.

To allow the hydraulic fluid to enter the system, you need to place the hydraulic switch/circuit breaker into the on position. The electrical power is switched off from the solenoid valve.

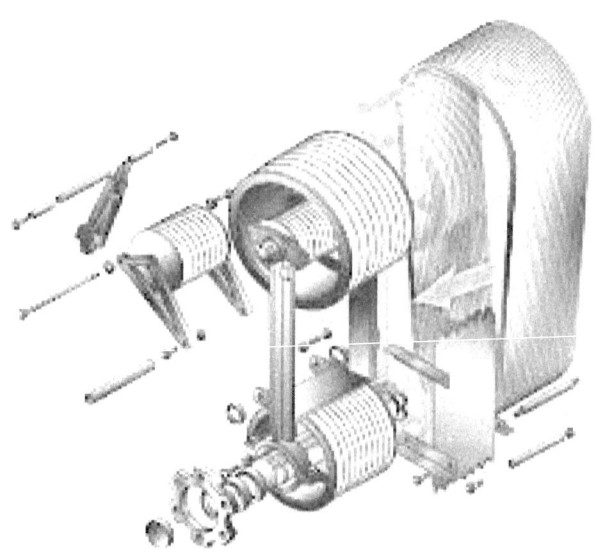

# Electrical Systems

Helicopters are equipped with power sources that provide either a direct current (DC) or alternating current (AC).

The power supply is at 28volt DC, with a single-conductor system that uses the helicopter structure as a  ground negative. The alternator provides the power to the DC system components like the starter, battery, or other external power.

A standby starter-alternator provides Emergency DC power with a switch, which is mounted on the engine. There are two inverters (main and spare) that convert the 28VDC to 115v  AC Supply of the alternating current. The inverters are identical, and both capable of supplying power for the operation of all AC-operated equipment.

In most helicopters, the electrical systems reflect the use of avionics and other electrical accessories. More operations in today's flight environment are dependent on the aircraft's electrical system. All helicopters can safely fly without electrical power in the event of an electrical malfunction or emergency. Like a vehicle, an engine-driven alternator supplies the electrical energy on a small, piston--power-engine helicopter.

The alternators have been better than the older style alternators because they light in weight and maintain a uniform electrical output even at low engine speed. When it comes to electricity, think of volts or voltage just like a water system.

The basic electrical quantities are electrical current and voltage, electrical charge, resistance, capacitance, inductance, and electric power. Electrical pressure is comparable to pounds per square inch.

To find the amount of energy consumed, you multiply the energy consumption (measured in watts) by the amount of time (measured in hours) that it is being consumed. Electrical energy is measured in watt-hours. Another way to think about power and energy is with an analogy to travel.

Each time you start the engine, the battery supplies electrical power to the starter/alternator to start the engine. Once the engine starts, the starter/alternator is used as the alternator. The alternator voltage is supplied via a voltage regulator.

The voltage regulator maintains the electrical system's constant voltage by regulating the alternator's output or the alternator bus bar.

An overvoltage control is integrated into the system to avoid excessive voltage, damaging the electrical devices.

The bus bar distributes power to the various electrical devices in the helicopter. A battery is the power source when starting the engine. It also allows the operation of radios and lights, without running the engine. In the event of an alternator failure, the battery can also provide backup power.

The electrical current in the system uses an ammeter or a charge meter to monitor the system. The ammeter displays the current flow to and from the battery. After an engine starts, the alternator charges the battery.

While the engine running, the alternator delivers the electricity to the system. When the electrical load exceeds the alternator's output, it means the battery helps supply the electrical power. It's because the alternator has a malfunction or excessive electrical load.

A load meter will display the load on the Alternator by the electrical equipment. The loss of the alternator causes the load meter to indicate zero. Electrical switches are built-in to select electrical apparatuses and provide power to the components.

The power may be supplied directly to the component by a relay. Heavy electrical cables that require particular components must use Relays with high current to protect the different electrical apparatuses from overloading and uses circuit breakers or fuses. When a respective component is overloaded, its circuit breaker pops out. You could reset the circuit breaker by switching off the breaker unless a short or the overload still exists. If the circuit breaker continues to pop, it indicates an electrical malfunction. When a fuse is overloaded and burn, you need to replace it. If one of the fuses has to replace in flight, the manufacturers usually provide a holder with a spare fuse. A malfunction of an electrical component could also set the warning light on the instrument panel.

# FUEL SYSTEM

# Fuel Supply System

The fuel system in a helicopter contains two systems: the fuel supply system and the engine fuel control.

The fuel system consists of a fuel tank/s, fuel pumps, gauges, a shut-off valve, and fuel filter. The fuel tanks are secured to the airframe as close as possible to the center of gravity (CG). As the fuel burned off, the effect on the CG insignificant. There is a drain valve on the bottom of the fuel tank that allows you to drain water and foreign deposits accumulated in the tank.

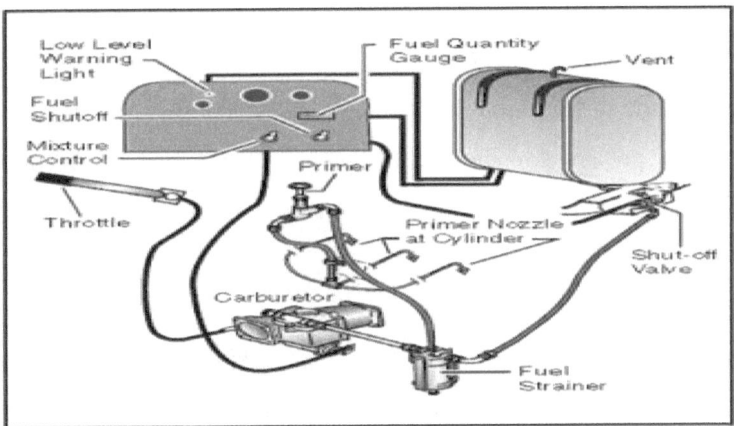

A fuel filter removes the moisture and other deposit from the fuel. The fuel vent prevents the formation of a vacuum in the tank. Fuel allows expansion without rupturing the tank by the overflow drain.

The fuel travels through a shut-off valve in the event of an emergency to stop the fuel flow.

The engine-driven pump is the primary pump that supplies fuel to the engine and operates any time the engine running.

An electric pump and a mechanical engine-driven pump are feeding the fuel systems. The electric fuel pump maintains a positive fuel pressure to the engine pump and serves as a backup in mechanical pump failure. A switch in the cockpit controls it.

A fuel quantity-gauge on the helicopter pilot's instrument panel shows the amount of fuel in the tank. Most fuel gauges must be accurate only when 'empty' to indicate the correct amount of fuel.

Title 14 CFR section 27.1337 states that the fuel quantity indicators 'must be calibrated to read 'zero' during a level flight when the quantity of fuel remaining in the tank equals the unusable fuel supply.

It always is a good routine to visually verify the fuel onboard and determine the amount before a flight and make sure an adequate fuel is present for the flight duration. Title 14 CFR section 27.1305 requires newer helicopters to warn systems to warn the flight crew when approximately 10 minutes of usable fuel remained in the tank. The systems must be calibrated, and never assume that the entire amount is available. Many helicopter pilots have not reached their destinations due to poor fuel planning or faulty fuel indications.

You should keep the needle above the yellow arc marker or within the green arc. If ice is allowed to form inside a carburetor in a carburetor engine system, an engine failure is a real possibility and may impair the ability to restart the engine.

With the fuel injection system, the fuel and the air all metered at the fuel control unit but are separated.

The fuel is directly injected into the intake port and mixed with the air, just before entering the cylinder. It's done so to ensure the fuel distribution is even though all the cylinders. The fuel injection system eliminates carburetor icing problems and the need for a heating system in the carburetor.

# Minimum fuel reserve

The minimum reserve fuel on-board the helicopter at the time of landing is about 300 lbs. This is designed as an absolute minimum and not to use for landing. If landing with less than the specified quantity of fuel on board, the pilot must submit a written report explaining the use of the reserve fuel specified within five working days.

# Fuel System Components

Under paragraph 14 CFR part 23, airworthiness standards specified for air carrier and helicopter certifications are similar.

The fuel used to power aircraft is a specialized type of petroleum-based fuel called Aviation fuel. Depending on the type and make of helicopters, there are several variations. A small two-seater-training helicopter with a piston engine pivotally burns between 9 to 16 gallons per hour. A larger five-seater turbine helicopter pivotally burns about 25 to 30 gallons per hour. The helicopter's normal cruising speed depended on the amount of power available and the type of its rotor system.

The two-seat style helicopter's pivotal cruise speed is about 90-105Mph and the five-seat turbine about 130-145 Mph. As far as maximum speed, 155Mph in a Bell 206 is common. Helicopter manufacturers generating flight performance data to estimate fuel needs for a flight. The fuel burn-rate is based on a specific throttle setting in flight.

# Aircraft Refueling Procedures

You must ensure the proper grade, quality, and desired amount delivered to the fuel tank. You may delegate the fueling of a helicopter, but as a pilot, you always retain the responsibility. You should verify that the refueling equipment is labeled with the name of the product ordered (Jet A, Jet B, etc.) and properly positioned and not under any part of the helicopter that could settle down during refueling.

In case of emergency, all the refueling trucks should approach in reverse to the aircraft, so they can quickly move away.

One helicopter pilot will supervise the refueling and attach the truck's nozzle ground wire before opening the fuel cap. When their passengers on-board the helicopter, they will have to come out before refueling executed. Refueling with passengers onboard may be refueled if the refueling cannot obtain when requested or in some special cases like injured passengers.

After refueling, the supervising crewmember will disconnect all the cables and pipes from the aircraft and physically check the fuel caps for security.

Hot fueling can be defined as fueling with the main aircraft engine running. When fueling helicopters with engines and blades turning, a person who uses a rapid fueling must have experience. A fire extinguisher must be readily available during all fueling operations, and a helicopter pilot must remain at the flight controls when the engine is running at idle.

This operation should not be considered a routine task. It should be done with extreme care, and only when necessary. It is a good practice that a written agreement is in place in advance of this type of operation, stating all areas of responsibility and liability of training requirements.

If a thunderstorm near the fueling facility, no fueling is allowed. During fueling, no radio transmissions shall be made, and all strobe lights should be turned off. The flight crewmember can request a sample taken from the fuel before accepting fuel into the helicopter if the fuel's quality or the fueling equipment in question.

The Fuel samples should have no evidence of suspended (cloudy or hazy) water, have a clear and bright appearance, and have no visible particulate matter (dirt, rust, etc.)

Unclean Fuel must free from water and should not be accepted.

# HELICOPTER CONTROLS

# Active Augmentation Systems

So-called actual augmentation systems use electric sensors and actuators that provided input to the hydraulic servos. These servos receive control commands from the computer that senses the external environmental inputs, such as wind or turbulence.

The hazards associated with helicopter flight in Degraded Visual Environments (DVE) have caused several military operations and civilian operations with an inadvertent flight into Instrument Meteorological Conditions by losing situational awareness.

This is a result of degraded visual conditions being significant contributors. It's a major contributor to the high accident rate for small helicopters and comes from an excessive pilot workload.

SAS system complexity is varied by manufacturers, but it's as sophisticated as providing three-axis stability. Computer-based inputs can adjust altitude, power, and aircraft trim for a more stabilizing flight.

# The Throttle

The throttle controls the power produced by the engine. The throttle is fixed at the end of the Collective, while in a multi-engine helicopter, it has power levers. Opening the throttle increases RPM. The throttle maintains engine power, so the rotor creates the lift necessary for flight while keeping the rotor speed within limits.

In most piston engines-power helicopters, you manipulate the throttle to maintain the rotor speed.

When the Collective is raised, the power automatically increases. When the Collective is lowered, the power is decreased to ensure the rotor's speed remains constant as the angle of attack changes.

If the main rotor and tail rotor not spinning fast enough, they cannot create enough lift, and the result could be catastrophic. Without the main rotor lift, a helicopter cannot stay aloft; without a tail rotor lift, you cannot maintain yaw control.

When the helicopter moves to the right due to the combination of main rotor torque and tail-rotor anti-torque, the forces to the right are greater. The helicopter tends to drift to the right, assuming no pilot inputs.

Translating tendency is called by many - a rotor drift.

# The Cyclic

During the flight, the Cyclic (short for 'Cyclic pitch control') controls the helicopter's pitch and the bank altitude like the yoke or stick that controls the elevator and ailerons in an aircraft. The Cyclic is the primary airspeed control. Directing forward the Cyclic will cause the airspeed to increase, and aft the cyclic decreases the airspeed. The Cyclic can pivot in all directions. The pitch control is fixed vertically between the helicopter pilot's knees or in the helicopter's center.

The Cyclic controls the direction and speed of the helicopter's movement across the ground by tilting the rotor disk in a hover flight. As the Cyclic moved forward, it flies the helicopter forward. Moving the Cyclic left makes the helicopter translating tendency, moves over the ground, to the left, and so forth. The Cyclic position determines the tilt (or direction) of the thrust vector. Applying aft Cyclic will cause the helicopter's nose to rise.

The range you move the Cyclic determines how fast the helicopter moves in a particular direction. Moving the Cyclic usually requires you to make adjustments with the Collective and the ant rotating force pedals. Here is an instance:

In normal cruising flight, when you apply forward the cyclic, the helicopter's nose will drop. Unless you increase the collective to increase the lift, the airspeed will increase, and the helicopter will descend.

Applying back the cyclic causes the helicopter's nose to rise. So, as the airspeed decreases, the helicopter will ascend, unless you apply it to reduce power.

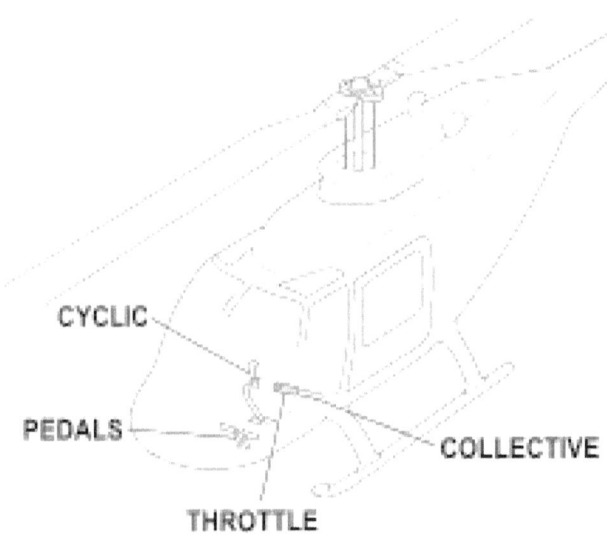

# Cyclic Pitch Control

The cyclic controls the rotor disk tilt against the horizon. Moving the cyclic forward, tilts the rotor disk forward. By moving the cyclic aft, the rotor disk tilts aft.

The cyclic pitch control tilts the plane in the direction of flight.

The cyclic pitch control pivotally raised up from the cockpit floor, in between the helicopter pilot's legs or, in some models, is between the two helicopter pilot seats. This flight control allows you to fly the helicopter in any direction.

The Anti-rotating force pedals can compensate for any change in the rotating force in a hover to control the heading.

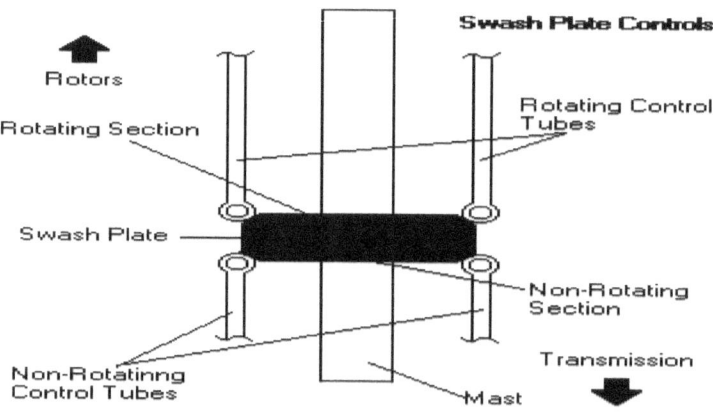

# Collective pitch control

The Collective serves as a helicopter's primary altitude and power control. It changes the pitch angle of all the main rotor blades collectively at the same time, and independent of their position. It's located on the left side of the pilot's seat with an adjustable friction control to prevent inadvertent movement. As you decrease the pitch angle, the angle of attack and the drag force will decrease, too, except that the rotor speed will increase. Use your left arm to raise and lower the Collective by moving the long lever mounted on the cockpit floor.

If you raise the collective, it will increase the angle of attack of all of the rotor blades at the same time and increase the lift. If you lower the collective, you decrease the angle of attack of all the blades simultaneously, thereby reducing the main rotor's amount of lift.

When you raise the Collective, the rotor blades produce more lift. Simultaneously, the increased angle of attack also produces more of the drag force, so you must increase power to maintain the rotor speed. The increase in power will cause an equal and opposite reaction by increasing the rotating force. Therefore, when you raise the Collective, you must also apply the left anti-rotational force pedal.

Lowering the Collective will decrease the lift and drag forces, and does not require additional power to maintain the rotor speed.

To maintain flight coordination, you must apply the right pedal pressure as you lower the Collective. If you wait to experience the effect of a controlled movement and then react, you will have trouble controlling the helicopter. In a straight flight, a helicopter might climb up or descent down.

The engine drives the main rotor and the tail rotor. Suppose you must use a high-power setting to maintain a hover in a strong crosswind. In that case, the tail rotor may not develop sufficient thrust to counteract the main rotor's rotating force, and the helicopter will tend to weather vane into the wind.

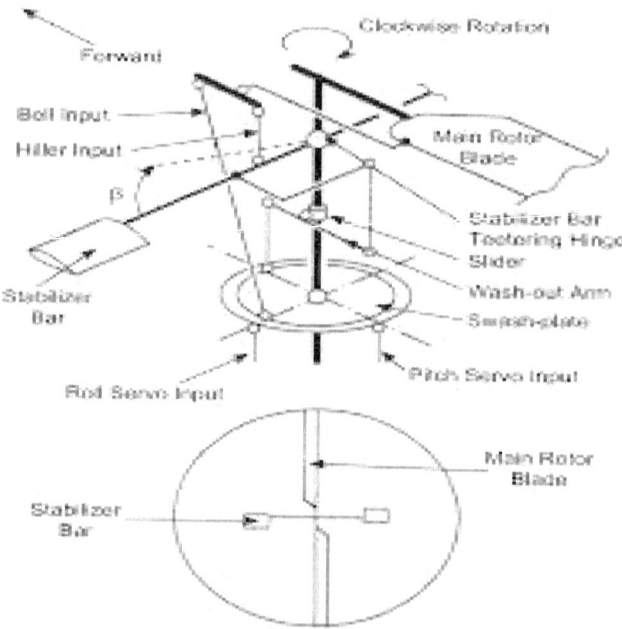

# Anti-torque pedals

The anti-torque pedals let the pilot control the tail rotor blades' pitch angle, which is a straight forward flight, puts the helicopter in longitudinal trim, while at a hover, and enables the pilot to turn the helicopter 360°. The pedals are located in the same position as the rudder pedals in an airplane and serve a similar purpose, namely to control the direction in which the helicopter's nose be pointed.

The pedals change the pitch of the tail rotor and altering the amount of thrust produced. The main rotor produced rotation force that needs compensating by using the anti-rotational force pedals.

Increasing Collective will cause more rotating force; decreasing Collective will cause less rotating force. You must use the pedals to avoid spinning out of control.

When you add power by increasing Collective, you must use a left pedal to keep the helicopter from rotating to the right. Likewise, if you decrease power by lowering the Collective, you must use the right pedal to compensate for the reduction in rotating force. A helicopter turns like an aircraft in straight forward flight, by banking. You can also practice for a left pedal to turn left or right pedal to turn right.

In cruising flight and during normal climbs or descents, you should use the pedals to maintain flight coordination and keep the helicopter in a narrow trim.

It is not recommended using pedals to turn, except in a hover. Instead, you may want to use the cyclic to bank and turn the helicopter. To recover from an unusual altitude, you can correct the pitch and the bank altitude by adjusting the power and trims the helicopter as necessary.

The bank altitude is corrected related to the turn and the banking indicator, and when the ball is centered, the airspeed checked and corrected and more power if required. If the ball is positioned left of the center, you should apply the left pedal or the right pedal pressure if the ball to the center's right. In most helicopters, the anti-rotating force of the rotor or tail rotor can be compensated this way.

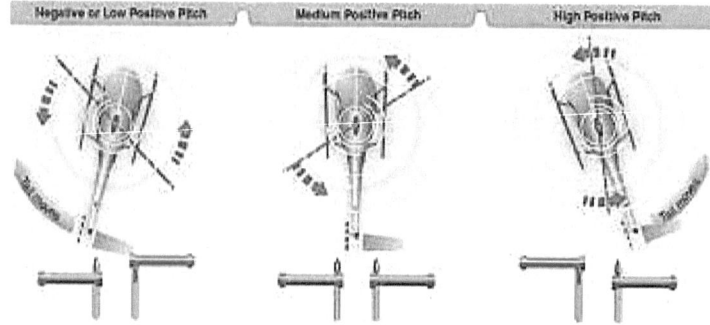

# Governor/Correlator

On all turbine helicopters, the governors common as part of the turbine engine's fuel control system. To keep the rotor speed constant, the governor makes the necessary adjustments as it senses the rotor and the engine speed.

Once the rotor speed is set in normal operation, the governor will keep the speed constant, and there is no need to make any more throttle adjustments.

, Between the Collective lever and the engine throttle, there's a mechanical connection called - Correlator. When you raise the collective lever, the Correlator increases the power automatically. Moreover, when you lower it, it decreases power. This system's actual speed value is maintained to match the desired value, but it still requires the adjustment of the throttle for a fine-tuning. The twist grip throttle is usually fixed on the end of the Collective lever.

On some turbine helicopters, the throttles are fixed on the overhead panel or on the cockpit floor.

Some helicopters do not have correlators or governors, which require you to coordinate the collective and the throttle movement.

As the Collective is raised, the throttle must be also increased, and is decreasing when the Collective is lowered. Large adjustments of Collective pitch or throttle should be avoided, as with any control. All corrections should be made as a smooth operation.

The Collective pitch is the primary control of the manifold pressure in piston helicopters, and the throttle is the primary control of engine speed.

The Collective pitch control also influences the engine speed, and since the pressure throttle influences the manifold, they are secondary control of each other's function. You must examine the tachometer (RPM indicator) and the manifold pressure gauge to govern which controller to use.

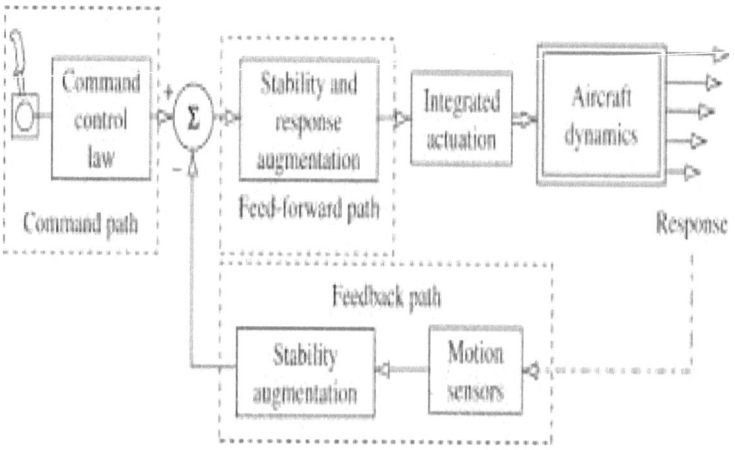

# Stability Augmentations Systems

Some helicopters incorporate a stability augmentation system (SAS) to stabilize the helicopter in flight and in a hover. The original purpose and design are allowing the pilot workload to decrease and lessened his fatigue. It allowed pilots to place an aircraft at a set altitude to accomplish other tasks or simply stabilize the helicopter for a long cross-country flight.

# Force Trim

Force trim is a passive system that simply holding the cyclic in a position that gave control to a  pilot who became accustomed to such control forces. The system uses a magnetic clutch and springs to hold the cyclic control in the position where it was released. The system does not use sensor-based data to make corrections but rather used by the pilot to "hold" the cyclic in the desired position. The most basic versions apply only to the cyclic and requiring the pilot to continue power and tail rotor inputs. With the force trim on or in use, the pilot can override the system by disengaging the system using a force trim release button or, with greater resistance, can physically manipulate the controls. Some of the recent basic systems are referred to as altitude retention systems.

# Heading Control

Heading Control is used to control the helicopter's heading while hovering or when making hovering turns and counteracting the main rotor's torque. The hovering turn referred to as pedal-Turn. The pedals are used to compensate for the rotating force at speeds above translational lift to put the helicopter in longitudinal trim.

By making a turn with the cyclic control in the desired direction, the heading change is achieved. The tail rotor's thrust governs the pitch angle of rotor blades.

A positive pitch angle inclines to move the tail to the right while a negative pitch angle moves the tail to the left. A zero-pitch angle produced with no thrust and the negative pitch angle will move the tail to the left. All in all, the rotor's maximum tail positive pitch angle is greater than the maximum negative pitch available.

The main purpose of the tail rotor to counteract the rotating force of the main rotor. From the neutral position, applying the right pedal will cause the helicopter's nose to yaw right and the tail to swing to the left. Pressing the left pedal has the opposite effect: the helicopter's nose will yaw to the left, and the trail swings to the right.

With the pedals in a neutral position, the tail rotors maintain a medium positive pitch angle. The tail rotor thrust is equal to the rotating force in this angle to the main rotor so that the helicopter will maintain a persistent direction in a cruise flight.

Helicopters with tandem rotors do not have an Anti-rotating force rotor. It is designed so that both of the rotors are rotating in opposite directions to counteract the rotating force.

# Autopiloting

A three-axis autopilot has an additional servo connected to the antitorque pedals and controls the helicopter in yaw. A four-axis system uses a fourth servo that controls the collective. These servos move the respective flight controls when they receive control commands from a central computer.

This computer can receive data input from the flight instruments for altitude reference and navigation equipment for navigation and tracking reference. Autopilot has a control panel in the cockpit that allows the pilot to select the desired functions and engage the autopilot. An automatic disengagement feature is usually included for safety purposes, which automatically disconnects the autopilot in heavy turbulence or when extreme flight altitude is reached.

Even though all autopilots could be overridden by the pilot, an autopilot disengagement button is located on the cyclic or collective, allowing pilots to completely disengage the autopilot without removing their hands from the controls.

Because the autopilot systems and their installations differ from one helicopter to another, it's very important to refer to the autopilot operating procedures located in the RFM.

# Centrifugal forces

Helicopter rotor systems rely primarily on rotation to produce a relative wind, which develops the aerodynamic force required for the flight. Due to its rotation and weight, the rotor system is exposed to Special Forces and moments with all kinds of rotating masses. One of them is the centrifugal force. This force is tending to move the rotating body away from the center of rotation. Another force that is generated in the rotor system the centripetal force. This force counteracts the centrifugal force by holding an object within a certain radius of the rotation axis.

The rotor blades of a helicopter generate very high centrifugal forces on the rotor head and the blade mounting assemblies. The centrifugal loads at the root of the blades of helicopters can be 6 to 12 tons. Larger helicopters can apply up to 40 tons of centrifugal force to each leaf root. In a rotary-wing aircraft, the centrifugal force is the dominant force that affects the rotor system. All other forces modify this force.

When the rotor blades at rest, they collapse due to their weight and range. In fully mobile systems, they rest on a static stop or vibration that prevents the blade from sinking so low that it could touch the helicopter. As the rotor starts to rotate, the blade starts to lift from the static position due to the centrifugal force. At operating speed, the blades extend to a straight position, even if they not flat and do not generate lift.

If the helicopter takes off, the blades rise above the "straight" position and assume a conical position. The cone's height depends on the G-forces during the flight, the speed, and the gross weight. If the speed is kept constant, the cone increases as the gross weight and the force G increase. If the gross weight and G-forces are constant, a decrease in speed will increase transformation.

# Translational lift

The improvement in rotor efficiency is resulting from a directional flight known as a - translational lift. The efficiency of the hover rotor system is greatly improved at each arrival node.

The combined upward beat (reduced lift) of the advancing blade and the downward beat (increased lift) of the receding blade will balance the lift across the main rotor disc and counteract the lift's asymmetry.

Efficient translational lift or ETL is a transition state present after a helicopter has switched from hovering to a forward flight. When a helicopter is hovering, the tail rotor operates in a highly disturbed airflow. When the helicopter reaches the ETL, the tail rotor starts to generate much more thrust force due to the less disturbed airflow. The helicopter will react to increase the yaw thrust.

You must reduce the tail rotor's thrust by pressing the pedal at about the same time as making the cyclic adjustment for a lateral tracking, acceleration, and climb.

Lateral flight can be a very unstable condition due to the parasitic drag force combined with a horizontal stabilizer for this flight path. Higher altitudes will assist in the control, and you must always sweep in the direction of flight.

# Ant rotating force

Most single main rotor system helicopters require a separate rotor to overcome the rotational force. This is accomplished with a variable pitch rotor and a rotary force or the tail rotor.

You will need to vary the ant rotation force system's thrust to maintain directional control whenever the main rotor's rotational forces are changing or make a course change in hover. The fenestra or 'fan-in-tail' design is another form of the ant rotating force rotor. This assembly is using a series of rotating blades mask within the upright tail. They are less likely to be exposed to people or objects because the blades are located within the circular duct. To maintain directional control, approximately 66% of the helicopter lift is necessary. The rest is created by the thrust from the rotating nozzle.

# Anti-torque rotor

An anti-torque rotor is located at the end of a tail boom extension. It provides compensation for torque in a single rotor helicopter. The tail rotor is driven at a constant speed. It produces thrust horizontally to oppose the torque reaction developed by the main rotor.

# Tail Rotor

Tail rotors are a smaller rotor/s, fixed in an upright position or near upright on a traditional rotor helicopter's tail. The counteraction for this rotating force is to push or pull the tail rotors against the tail. The tail rotor's drive system consists of a drive shaft that is power-driven by the main transmission, and the gearbox, which mounted at the end of the tail boom.

The driveshaft is a long shaft or shafts connected at the end with joints. The joints allow the driveshaft to bend or stretch with the tail boom.

The built-in gearboxes improve the power necessary to tilt the tail rotor drive-shaft and use it as a vertical stabilizing airfoil on some larger helicopters. The pylon (or vertical fin) provides a partial ant rotating force if the tail rotor or its flight controls is failing.

# Tail Rotor Aerodynamics

Although Helicopters come in a variety of sizes and shapes, they share the same major apparatuses. Tail rotors share many of the aerodynamics with the helicopter main rotor system. They identical to the main rotor, but they mounted sideways and controllable by the collective pitch. The tail rotors are not capable of a cyclic feathering. The tail rotors share the same problems and solutions as those of the main rotors. Since the tail rotor blades are allowed to flap, the cyclic pitch's use to counter the rolling tendency eliminates the lift's dissymmetry.

# NOTAR system

NOTAR is the name of a helicopter system, which replaces the use of the tail rotor. The name is derived from the phrase - no tail rotor. The NOTAR system is simple and works to provide some directional control. It is bounded in the aft fuselage sector, immediately forward of the tail boom, and has a variable pitch fan driven by the main rotor transmission.

This fan is forcing air through two slots on the right side of the tail boom. It then causes a downwash from the main rotor and produces lift, so to provide directional control.

At the end of the tail boom, a rotating drum provides a directional yaw control, gained through those vents. The system offers quieter and safer operation.

The advantages of the NOTAR system have increased flight safety and considerably reduced the outside noise.

The NOTAR system is not as efficient as the tail rotor system, and therefore the drawback of the NOTAR helicopter is the lack of power.

The NOTAR system eliminates all of the tail rotors' drawbacks, counting the long drive shafts, the bearings, intermediate gearboxes, and the ninety-degree gearboxes. 20% of all crashes result in a tail rotor striking or the loss of tail rotor efficiency, the NOTAR systems eliminating these glitches.

NOTAR equipped helicopters designed to be 50 percent quieter than other helicopters. They are among the quietest helicopters available today.

The phenomenon in which a flow of air-jet attaches itself to a nearby surface remains there even when the surface curves away are called - Coanda Effect. In other words, a jet of fluid mixed with its surroundings as it flows away from a nozzle.

In-flight, the vertical stabilizers are providing the major part of anti-torque. The directional control is remaining the function of the direct jet thruster.

By using the Coanda Effect, it could provide anti-torque to the tail boom and create a lift. The system is stable and much easier to control. This Coanda Effect is reducing the sensitivity to the wind direction.

# Rotor Safety

The nature of the main and the tail rotors deserves special attention. When taxiing near hangars or obstructions, you must exercise caution since it's very difficult to judge the exact distance between the rotor blade tips and any obstructions. On some helicopters, the tail rotor cannot be seen from the cabin.

When the passengers are approaching the helicopter on a slope with the rotors turning, they should approach and depart downhill. This affords the greatest distance between the rotor blades and the ground. If this must involve walking around the helicopter, they should always go around the front and never from the rear.

# Tandem helicopters

Tandem rotor (dual rotor) helicopters have two large horizontal rotor assemblies, with a twin-rotor system instead of one main assembly, and a smaller tail rotor.

Single rotor helicopters need an anti-rotating force system to neutralize the twisting momentum produced by the large rotor.

Tandem rotor uses the counter-rotating rotors, each one is canceling the other's rotating force. Counter-rotating rotor blades do not collide and destroy each other as they flex into the other rotor's pathway.

This configuration also has the advantage of being able to hold more weight with shorter blades since there are two sets.

Besides, all of the engines' power can be used for lift, whereas a single rotor helicopter uses its power to counter the rotating force.

# FLIGHT PRINCIPLES

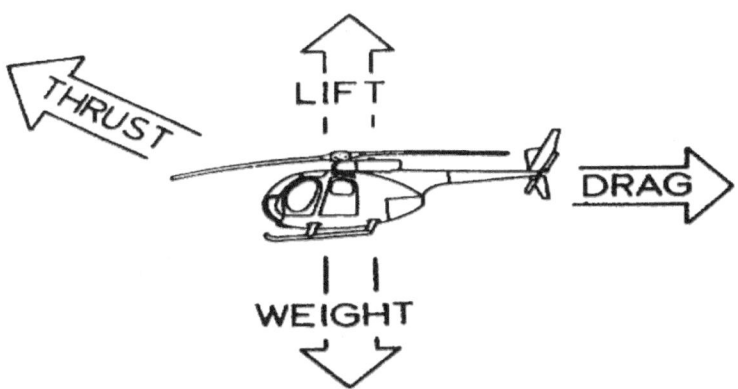

In un-accelerated forward flight, with a constant heading, and at a constant altitude, the lift equals the weight, and the thrust equals the drag force. If the lift force is greater than the weight factor, the helicopter will climb, and if the lift is lesser than the weight factor, the helicopter will descend. If the thrust exceeds the drag force, the helicopter will speed up. If thrust is lesser than the drag force, it will slow down. This basic principle of helicopters is not all that complicated. They fly by the same aerodynamic principles as any other aircraft. The force of Lift, Drag force, Thrust, and Weight are all needed to keep an aircraft in the air. The wings' shape designed, so the airflow over the top surface travels further and faster than the air under the wing. As the air pressure on top of the wing is reduced, the aircraft get sucked up by this lower pressure. The Lift is produced independently of the power from the engine.

The other force involved is 'Weight.' The aircraft's weight is defining how much lift is required to get it into the air. As long as there is enough lift to overcome the aircraft's weight and enough thrust from the engine to overcome the drag force, the aircraft will fly. Weight and drag forces are acting on a helicopter as they do on airplanes.

The lift and thrust of a helicopter are obtained from the main rotor. By tilting the main rotor, you can fly the helicopter from one side or the other. There are four basic flight principles, on which all maneuvers are based on straight turns, climbs, and descents.

All controlled flight maneuvers consist of one or more of the four basic principles of flight.

# The Lift force

Lift occurs when an object changes the direction of fluid to flow, or when a passing object forces the fluid to move. When an object and a fluid move relative to each other and the object rotates the fluid flow in a direction perpendicular to the fluid, the required force produces an equal and opposite force, namely, a lift force.

Objects can flow through stationary fluids, or fluids can flow through stationary objects; the two actually are the same because, in principle, only the audience's frame of reference is different. The lift generated by the aerodynamic curve depends on the following factors:

• Airflow speed

• Air density

The total area of a segment or aerodynamic profile

• The Angle of Attack (AOA) between air and aerodynamic curve AOA the angle at which the aerodynamic curve meets the oncoming airflow (and vice versa). In a helicopter, the targets are the rotor blades (aerodynamic profile), and the fluid is in the air.

Lift occurs when the air mass is deflected, and the lift is always vertical to the produced relative wind.

The asymmetric aerodynamic profile must have a positive AOA to produce a positive lift. When AOA equal to zero, no lift is generated.

When AOA is negative, the negative lift will be generated. A curved or asymmetric airfoil can generate a positive lift at zero or even a small negative AOA.

The basic concept of an elevator is simple. However, how air and airfoils' relative motion interact to produce a rotational action that generates lift are complex. The area within the circle formed by the ends of the helicopter's rotating blades is called - the disc or - the rotor disc area.

When hovering in calm air, the lift is generated equally by the rotor blades to all disc area parts.

Lift asymmetry refers to the difference in lift between the first half and the second half of the disc area caused by the horizontal flight or the wind.

When the helicopter still hovering in the air, the maximum speed of the advancing blades is about 600 feet per second.

The maximum speed of the concave blade is the same.

The Lift asymmetry is caused by the helicopter's horizontal movement during the transfer flight, while the blades are advanced at the combined speed of blade speed plus helicopter speed.

# Lift Mass

Periodic feathering will change the rotor disk's altitude, but will not change the rotor system's lift. Most changes in AOA come from changes in climb or descent speed and the speed itself.

Due to the design of the rotor system, the beats occur automatically. The beat is the up or down movement of the rotor blades.

Helicopter pilots could adjust the AOA through a normal control manipulation of the blade pitch angle. If the elevation angle increases, the AOA will increase; if the elevation angle decreases, the AOA will also decrease.

# Parasitic Drag force

Each time the helicopter flies in the air, parasitic resistance occurs. This medication increases with speed. Helicopter non-elevators, such as cabins, rotor masts, tail, and landing gear, can cause parasitic resistance.

Due to engine operation, cooling, engine load, and so on, there is a momentum loss, causing additional parasitic resistance. Because it increases rapidly with the increasing speed, the parasitic resistance is the main reason for resistance. The parasite resistance is always proportional to the square of the speed. Therefore, doubling the speed quadruples the resistance of the parasite.

# Thrust

Thrust, like lift, is generated by the rotation of the main rotor system. In a helicopter, the thrust could be forwarded, rearward, sideward, or vertical. However, the engine produces the thrust, which gets the aircraft moving along the runway to flow over the wings. Thrust also overcomes the Drag force, which is the force, which opposes the motion of the aircraft through the air. The resultant lift and thrust will determine which direction the helicopter will move.

The amount of thrust varies with the ant rotating force pedals use to control the helicopter's yaw.

When the task is not performed flawlessly, or you attempt one of the maneuvers that require more lift than the rotor system can produce, more power than the helicopter's power plant could provide, or try landing with the helicopter's nose at a high altitude, and most likely will end in a disaster.

# Total Drag force

For a helicopter, the sum of all three forces, the drag force is the worst. As the airspeed increase, the parasitic drag force increase too. If you look at a helicopter's drag force profile, it remains relatively constant throughout most conditions, but at higher airspeeds, it increases.

Combining all drag forces results in a total drag curve. The low point on the total drag curve shows the airspeed at which drag is minimized—this an important factor for the helicopter's optimum performance.

The lift-to-drag ratio, or L/D ratio, is the amount of lift generated by a wing or vehicle, divided by the aerodynamic drag it creates by moving through the air.

# The Venturi effect

Airflow always accelerating over the top of the surface of the airfoil under the Venturi-effect rule. As air velocity will increase as the air moves through a constricted portion of a Venturi tube, its pressure decreases.

The upper surface

Can compare with this constriction in a Venturi tube, narrower in the middle than at the ends. If the layers of undisturbed air replace the upper half of the Venturi tube, then when the air flows over the airfoil's upper surface, the airfoil's camber causes the airflow to increase in speed. Therefore, the increased speed of airflow causes pressure on the upper surface of the airfoil to decrease.

At the same time, airflows along the lower surface of the airfoil build up pressure.

The increased pressure on the lower surface and the decreased pressure on the upper surface results in an upward force. As the angle of attack increased, the lift increased. As the leading-edge stagnation point moves under the leading edge, a wash is created ahead of the airfoil.

Therefore, more downwash created aft of the trailing edge. The impact pressure and deflection of air from the rotor blades' lower surface provide a small percentage of the total lift. The majority of lift resulted from the decreased pressure above the blade, rather than the increased pressure below it.

# FLIGHT MANEUVERS.

There are two basic flight conditions for a helicopter-hover and forward flight. Hovering is the most challenging part of flying a helicopter. This because a helicopter generates its own gusty air while in a hover, which acts against the fuselage and flight control surfaces.

The result of constant control inputs and corrections is to keep the helicopter where it is required to be despite the task's complexity, the control inputs in a hover simple. The Cyclic used to eliminate drift in the horizontal direction. The Collective used to maintain altitude.

The pedals are used to control nose direction or the heading. The interaction of these controls makes hovering so difficult since an adjustment in one control requires an adjustment of the other and requires constant correction. It displaces the Cyclic forward, causing the nose to pitch down initially, with a resultant increase in airspeed and altitude loss.

Aft Cyclic causes the nose to pitch up initially, slowing the helicopter and causing it to climb; however, as the helicopter reaches a state of equilibrium, the horizontal stabilizer levels the helicopter airframe to minimize drag force, unlike a helicopter.

Coordinating these two inputs, down collective plus aft Cyclic or up Collective and forward Cyclic, results in airspeed changes while maintaining a constant altitude.

The pedals serve the same function in both a helicopter and a fixed-wing aircraft, to maintain balanced flight. This is done by applying a pedal input in whichever direction necessary to center the ball in the turn and bank indicator.

As an aircraft, the helicopter's primary advantages are the rotor blades that revolve through the air, providing lift without requiring the aircraft to move forward. This able the helicopter to take off and land vertically without the need for runways.

Thus, helicopters are often used in congested or isolated areas where fixed-wing aircraft cannot take off or land. The rotor lift also allows the helicopter to hover in one area and do so more efficiently than other vertical takeoff forms and landing aircraft, allowing it to accomplish tasks that fixed-wing aircraft cannot perform.

Helicopter piloting a helicopter requires a great deal of training and skill and continuous attention to the machine. You must think in three dimensions and use both arms and both legs constantly to keep the helicopter in the air.

Coordination, control touch, and timing all used at the same time when flying a helicopter.

Reliable helicopters capable of stable hover flight were developed decades after fixed-wing aircraft. Raising the Collective pitch control increases the pitch angle, or angle of incidence, by the same amount on all blades. The Collective pitch control called 'Collective' is located on the left side of the helicopter pilot's seat and operated with the left hand. The Collective used to make changes to the main rotor's-pitch angle and does this simultaneously, or collectively, as the name implies. As the Collective pitch control raised, there a simultaneous and equal increase in pitch angle of all main rotor blades; as it lowered, there a simultaneous and equal decrease in pitch angle.

This is done through a series of mechanical linkages, and the amount of movement of the Collective lever determines the amount of blade pitch change.

An adjustable friction control helps prevent inadvertent Collective pitch movement. Changing the pitch angle of the blades changes the angle of incidence on each blade. With a change in the angle of incidence comes a change in drag force, which affects the speed or revolutions per minute (SPEED) of the main rotor. As the pitch angle increases, the angle of incidence increases, drag force increases, and rotor speed decreases. Decreasing the pitch angle decreases both angles of incidence and drag force, while rotor speed increases. To maintain a constant rotor speed, which vital in helicopter operations, and a change in engine power required to compensate for the change in drag force. The engine speed (power) is adjusted automatically by the throttle control or the governor.

# Autorotation

The autorotation feature in a helicopter means that the main rotor system is turned by air moving up through the rotor rather than the engine power driving the rotor.

In normal, powered flight, the air is drawn into the main rotor system from above. It exits down, but during autorotation, and the helicopter descends, the air streaming upward from underneath the rotor system.

If engine power fails or certain other emergencies occur, the autorotation feature allows a helicopter's safe landing. When the engine stops, the transmission in a helicopter so designed to allow the main rotor to turn freely in its original direction. Autorotation is permitted mechanically by a freewheeling unit, in which a special clutch mechanism allows the main rotor to continue turning even if the engine is not running.

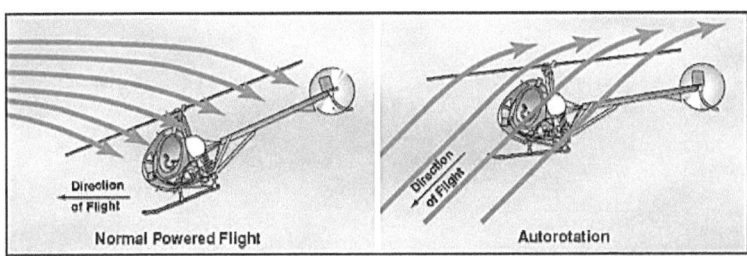

Direction of Flight

**Normal Powered Flight**

Direction of Flight

**Autorotation**

90

The freewheeling unit automatically disengages the engine from the main rotor. The rotor blade autorotative driving region the portion of the rotor blade that between 25 to 70 percent of the radius. All helicopters must demonstrate this capability to be certified.

If a decision is made to attempt an engine restart in flight, this emergency procedure different for each helicopter and must follow precisely. Once the engine started, the freewheeling unit will reengage the engine with the main rotor.

Autorotation equivalent to a power-off glide in an aircraft, and it helps to land after an engine failure.

During autorotation, it's important to maintain rotor speed, so it has a lift available to cushion the landing. You must also maintain the correct forward speed so that you can reach a suitable landing area and flare to reduce the rate of descent before ground contact.

# Autorotational regions

The rotor disc is divided into three regions during a vertical autorotation—those the driven, driving, and stall sections. These regions' size varies with the blade pitch, the rate of descent, and the rotor speed. The driven region at the end of the blades entails about 30 percent of the total radius. Most of the drag force produced by the driven region. The result of a slowing of the blade rotation speed.

The driving region, or authoritative region, normally ranges within the 25 to 70 percent of the blade disk, which produces the forces needed to turn the blades during autorotation. In the driving region, the total aerodynamic force inclined slightly forward, producing a continual acceleration force. This inclination supplies the thrust and accelerates the blade rotation.

The inner 25 percent of a rotor blade is the stall region. It operates above its stall angle, causing a drag force that slows the blade rotation. You can adjust autorotative speed by controlling the size of the driving region. The point of equilibrium then moves inboard along the blade span, increasing the driven region's size. As the stall region grows larger, the driving region getting smaller. Reducing the driving region's size will cause the driving region's acceleration force and speed to decrease.

A common cause for the autorotation is an engine malfunction or failure. Autorotation can also execute in the event of failure of a tail rotor, or in case the tail-rotor operation ineffective.

Since there is no rotating force produced in an autorotation, it may also recover from a vortex state if altitude permits alterations.

In any case, a successful landing depends on the helicopter's height and its velocity at the beginning of autorotation. The main rotor blades will produce thrust and lift within their attack and velocity angle at the instant of engine failure. When an engine fails to function, the Collective pitch must lower. If you immediately do that, you can reduce lift and drag force, and the helicopter will begin an immediate descent while producing an upward flow of air through the rotor system.

Since the tail rotor is powered by the main rotor transmission during autorotation, the heading control should maintain as in normal flight.

Several factors affect the rate of descent in autorotation: density altitude, gross weight, rotor speed, and forward airspeed. The helicopter pilot's main task is to control the rate of descent and airspeed. Higher or lower airspeeds are obtained with the cyclic pitch control just as in normal flight. At zero airspeeds, the rate of descent the highest, and it declines to a minimum of between 45 to 89 knots, depending upon the particular helicopter.

If the airspeed increases beyond the minimum rate of descent, the rate of descent will increase again. Even at zero airspeeds, the rotor is quite effective as it has the drag force factor. At this point, the kinetic energy stored in the rotating blades is used to decrease the rate of descent into a soft landing. A helicopter that descended more slowly required less rotor energy than a helicopter with a higher descent rate.

Therefore, authoritative descents at very low or very high airspeeds more critical than authoritative descents performed at the minimum rate of descent airspeed

# Descent

Each type of helicopter has specific airspeed at which a power-off glide is most efficient. The best landing airspeed the one that combines the greatest glide range with the slowest rate of descent.

Reducing the descent aerodynamic rate and slowing the groundspeed result in speed reduction down to approximately 15 feet. After that, the use of collective pitch further slows the descent and cushions the touchdown.

A helicopter operated with heavy loads in high-density altitude or gusty wind conditions can achieve the best performance from a slightly increased airspeed when descending. At low-density altitude and light loading, the best performance achieved from a slight decrease in normal airspeed.

You can achieve the same glide angle in any situation and estimate the touchdown point by following this general procedure of fitting the airspeed to the existing conditions of optimum glide angle of 17-20 degrees.

# Retreating Blade Stall

This situation occurs when a helicopter flies at high speed, and the angle of attack of the rotor blades move toward the rear of the aircraft. This occurrence is the primary limitation of helicopters' maximum speed. High-density altitude or a high gross weight can also induce a retreating blade stall.

In a retreating blade stall, the rotor blades on the rotor disk's left side exceed their critical angle of attack and stall. The sign of a retreating blade stall is a low-frequency vibration, and a tendency of the helicopter's nose to pitch up and a roll to the left.

To recover from a retreating blade stall, you must slow down and lower the collective to reduce the attack's rotor angle.

# Settling with Power

Most of the helicopters have a tendency to descend rapidly into their own rotor downwash. Settling with power is the most common cause of helicopter accidents, tied to a high descent rate.

Settling with power occurs when the helicopter airspeed less than 10 knots, during the descent of more than 300 feet per minute, and developing more than 20 percent power. Helicopter pilot encounters these conditions as you hover out of ground effect without holding the same altitude. When attempting to hover above the helicopter's out-of-ground-effect ceiling, which blows the downwash beneath the helicopter?

Vortices form at the root of the rotor blades and then travel outward. Like other stalled aircraft's wing, the rotors no longer produce enough lift to maintain the helicopter in level flight or maintain a gradual descent. It is usually associated with the cyclic loss of efficiency or a rapid descent rate.

To recover, you must move the helicopter out of its downwash by traveling forward, backward, or to the side at an airspeed of more than 10 knots.

# Flights Direction

Hovering is actually an element of a vertical flight. If you increase the rotor blades (pitch angle) while keeping their speed constant, the additional lift is generated. If you decrease the pitch angle, the helicopter rises and descends.

In unaccelerated forward flight, the lift equally heavy, and the thrust the same. Straight and straight flight means flight with a constant heading at a constant altitude. A lift, when the left greater than the weight it will cause the helicopter to climb. When the lift lesser than the weight, the helicopter will descend. When the thrust greater than the drag force, the helicopter will speed up. If thrust less than the drag force, the helicopter will slow down.

For the duration of hovering flight, the tip-path plane parallels to the ground. For the helicopter to hover, the sum of all the lift and thrust forces must equal the weight and drag force forces.

The tip-path plane the imaginary circular plane outlined by the rotor blade tips. In forward flight, the tip-path plane tilted forward, therefore tilting the total lift-thrust force forward. The tip-path plane tilted sideward in the direction that flight desired in sideward flight, thus tilting the total lift-thrust will vector sideward. The vertical or lift up, the weight straight down, and the thrust acting sideways while drag force acting in the opposite direction.

For the rearward flight, the tip-path plane tilted rearward, as the lift-thrust directed rearward.

The thrust rearward and drag forces directed forward, just the opposite of straight forward flight. The main rotor of a helicopter acting similar to a gyroscope. It has the same properties as a gyroscope move, and one of them is total precession.

When the gyroscope spinning, it can contain large amounts of stored energy. Due to gyroscope precession, the pitch change on the rotor blade doesn't occur where it should. The rotor's lift forces occur 90 degrees later in the rotation because of the gyroscope forces acting on the spinning rotor. This is a Gyroscopic Precession.

Gyroscope precession is a result of the spinning object when a force is applied to this object. Once the tip-path plane tilted forward, the total lift-thrust force also tilted forward.

As the helicopter begins to accelerate from a hover, the rotor system becomes more efficient due to translational lift. Continued acceleration causes an even larger increase in airflow, to a point, through the rotor disk and more excess power.

To maintain an un-accelerated flight, you must understand that the helicopter begins to climb or descend with any changes in power or cyclic movement. Once a straight-and-level flight is obtained, you should note the rotating force setting required and not make any major adjustments to the flight controls.

# Low-G Condition

A low-G situation develops whenever there is less than 1G on the rotor disk, which the helicopter's weight. You can induce a low-G situation by making abrupt Cyclic control inputs when pushing out from a steep climb or encountering turbulence. At this point, the nose will drop, and the aircraft may roll rapidly to the right because the tail rotor above the main rotor disk and thrust produced to the right. Helicopters with semi-rigid rotor systems may experience mast bumping as the main rotor hit the tail boom. In both cases, it may cause a loss of the main and the tail rotors.

You must gently apply aft Cyclic to raise the nose and add gas to the rotor disk before you lose control, trying to recover from a low-G situation.

# HELICOPTER PERFORMANCE

Helicopters are much more sensitive to control than most airplanes. To easily fly a helicopter and be precise, you must coordinate the use of all flight and power controls.

As you adjust the Cyclic to the side, the rotor disk will tilt to the same side, producing thrust in that direction and causing the helicopter to hover sideward. To eliminate drift in the horizontal direction, the Cyclic is used to control the forward and back, right and left. The Collective is used to maintain altitude. The pedals are used to control the nose direction or the heading of the helicopter. Since an adjustment in any one control requires an adjustment of the other two, to create a constant correction of the cycle, these controls' interaction makes hovering difficult.

When moving the Cyclic forward, it causes an initial tilt of the nose, which results in increased speed and loss of altitude. When the helicopter reaches a steady-state, the horizontal stabilizer levels the helicopter airframe to minimize the drag force, unlike an airplane. In a helicopter, the pedals perform the function of maintaining a balanced flight by applying a pedal in any direction necessary to center the ball on the turn and tilt indicator.

Consequently, the helicopter has very little upward or downward deflection when it is a stable in-flight mode. The variation depends on the particular helicopter and the horizontal stabilizer function.

The increase in the collective (power) while maintaining a constant speed causes an ascent, while the collective decrease causes a descent.

Anticipating how moving the control will correspond to movements from other controls. For instance, if you add power by increasing the Collective, you must also add a left pedal to compensate for the helicopter's tendency to rotate to the right.

You must understand that the special aerodynamic effects are unique to specific helicopters and the proper control inputs type needed to compensate for them. You must anticipate these effects, not just react to them. If you wait to experience the effect and then react, you will have trouble controlling the helicopter. You should never remove your hand from the Cyclic while the main rotor turning.

After landing, you should make sure the helicopter has firmly settled, and the Collective is in a totally down position as you prepare to shut down the engine. You should hold the Cyclic in the neutral position until the main rotor stops spinning.

Some helicopter pilots are practicing quick maneuvers of the ground operation to avoid airport terminals' delays and lessen the engine's stop or start phases.

You may leave the cockpit with the engine and rotors running. This operation could be extremely hazardous if a gust of wind disturbs the rotor disk or the Collective moves and causes a generated lift by the rotor system.

It may cause the helicopter to roll or pitch, and the rotor blade may strike the tail boom or the ground.

In general, a good operating procedure dictates that whenever the engine is running, and the rotors are turning, helicopter pilots must remain at the flight controls. If you need to leave the cockpit to refuel, the throttle should roll back to idle, and all controls must firmly held to prevent uncontrolled movements of the helicopter. You should be well trained in setting the controls and exiting the cockpit.

When the flight is terminated, you must park the helicopter in an area where it does not interfere with other aircraft and does not create a hazard for people during the shutdown procedure. It is advantageous for many helicopters to land with the wind coming from the right over the tail boom (counter-rotating blades). This tends to lift the blades over the tail boom but lowers the blades in front of the helicopter. This action decreases the likelihood of the main rotor strike to the tail boom due to gusty winds. Rotor downwash can damage other aircraft nearby, and spectators may not realize the danger or see the rotors turning. All passengers in the helicopter should remain seated with their seat belts secured until the rotors have stopped turning.

During shutdown and post-flight inspection, you must follow a checklist provided by the manufacturer. When discrepancies or issues are recognized, they should be noted, and if it is necessary, reported to the maintenance personnel.

# Autopilot

Helicopter autopilot systems are similar to the stability augmentation systems, but they have an additional feature. An autopilot can actually fly the helicopter and perform certain functions selected by the pilot. These functions depend on the type of autopilot and what systems are installed in the helicopter.

The most common functions are altitude and heading hold.

Transitioning helicopter pilots who were accustomed to control the force trim are holding the Cyclic in a position that gives them a passive control. This system holds the cyclic control in the same position where it was released by using a magnetic clutch with spring and is referred to as an - altitude retention system.

Those systems use many sensors from stabilized gyros to electro-mechanical actuators, which provide instantaneous inputs to all flight controls without the pilot's need.

These systems are very useful when performing other tasks, such as sling loading or search and rescue operations. The Autopilot systems are similar, but it has additional functionality. The autopilot can perform certain functions selected by the pilot and fly the helicopter, depending on the autopilot type and the helicopter's systems.

The most common functions are altitude and heading maintenance. Some more advanced systems include a vertical speed or an indicated speed standby mode (IAS). The constant rate of climb/descent or IAS is maintained by the autopilot.

Some autopilot systems have navigation capabilities, such as Omni Range of Very High Frequency (VHF). It also includes the navigational system (VOR), the instrument landing system (ILS), and the global positioning system (GPS) intercept and track, which is particularly useful under the conditions of instrument flight rules (IFR).

An additional component, called a flight director (FD), can also be installed. The FD provides visual guidance indications to you to pilot certain lateral and vertical modes of operation.

The most advanced autopilots can perform a hovering instrument approach without any additional pilot intervention once the initial functions have been selected. The autopilot system consists of electric actuators or servos, which are connected to the flight controls. A two-axis autopilot controls the helicopter in pitch and roll. One servo control is the front and rear of the Cyclic, and the other servo controls are the left and right cyclic.

A three-axis autopilot has an additional servo connected to the anti-rotation force pedals and controls the helicopter. There also four-axis systems, which use an additional servo that controls the collective. These servos move the respective flight commands when they receive control commands from a central computer. This computer receives input data from flight instruments for altitude reference and navigation equipment for navigation and tracking reference.

The autopilot has a control panel in the cockpit that allows you to select the desired function and engage the autopilot. For safety reasons, an automatic disengagement function is included, which automatically disconnects the autopilot in the event of heavy turbulence or when extreme flight altitudes are reached.

You can cancel the autopilots. The autopilot disengagement button is located on the cyclic or collective and allows you to disengage it without removing his hands from the controls.

Since autopilot systems and installations differ from helicopter to helicopter, it's very important to refer to the autopilot operating procedures located in the RFM.

# Cooling the helicopter

Environmental systems, heating, and cooling of the helicopter cabin can be achieved in different ways. Dynamic air is the simplest form of cooling. You control the air in the cabin by opening or closing air ducts in the helicopter.

This system is limited because it requires a forward speed to provide airflow and depends on the outside air temperature. Air conditioning provides better cooling, but it is more complex and weighs more than a dynamic air system.

Removing the doors and allowing air to circulate in the cockpit and engine compartments the simplest method of cooling the helicopter. Doors must properly store with care inside a hangar or, if necessary, be transported inside the helicopter.

When storing the doors, care must take not to scratch the windows. Particular attention should pay to the storage of all seat belt cushions and any other loose items to avoid ingestion into the main or rear rotor. When replacing the doors, care must be taken to ensure that they are completely secure and closed. Air conditioners or heat exchangers can also install on the helicopter.

They work by sucking the purge air from the compressor, passing it through the heat exchange, then releasing it into the cabin. When the compressed air is released, the expansion absorbs the heat and cools the cabin.

# A nonsymmetrical Airfoil

The nonsymmetrical airfoil has different upper and lower surfaces. They have a greater curvature of the airfoil above the chord line than below it. The nonsymmetrical airfoil is designed to produce useful lift at zero AOA, and its design has the advantages of more lift production than asymmetrical design. It also has better stall characteristics.

# Resultant Relative

The relative wind speed components are resulting from the helicopter moving through the air. This speed component is added to or subtracted from the relative wind of rotation, whether the blade moves forward or backward relative to the helicopter's movement. When the helicopter in a horizontal movement, the speed changes the resulting relative wind.

The resulting relative wind also serves as a reference for developing vectors of lift, drag force, and the total aerodynamic force on the aerodynamic profile.

The relative wind could result in hovering in a relative wind and modified by the inductive flow. It tilted downwards at a certain angle opposite the effective flight path of the aerodynamic profile.

The air circulation through the disc changes its pattern when the helicopter in a horizontal movement. As the helicopter gains speed, the addition of the transmission speed decreases the induced flow speed. This change is resulting in improved efficiency (additional lift) produced from a given blade pitch setting.

As the blades' tilt angle increases, the rotor system will induce a downward airflow through the rotor blades and create a downward air added to the relative rotational wind.

Because the blades move horizontally, part of the air is moving down. The blades move on the same path and pass a given point in rapid succession.

The action of the rotor blade is transforming the calm air into a column of the descending air. As a result, each slide has a reduced AOA due to backwashing. This downward airflow is called - induced flow (backwashing).

It is most pronounced when hovering in windless conditions.

# Velocity

The helicopter's main rotor blades must move through the air faster to produce sufficient lift to lift and hold the helicopter. The main rotor can rotate at the required take-off speed, while the anti-torque rotor keeps the fuselage speed at zero. The helicopter can fly forward, backward, or sideways as desired by the helicopter pilot. It can also remain stationary in the air (when hovering), as the main rotor blades are developing the lift to support the helicopter.

For helicopter rotor blades, the flight path's speed is equal to the speed of rotation, plus or minus a directional speed component. During the rotor rotation, the speed-of-blades' rotation is the lowest close to the hub and increases towards the blade's tip.

The Relative wind is created by the movement of an aerodynamic profile in the air. As an induced airflow can change the flight path's speed, the relative wind undergone by the airfoil may not be exactly the opposite to its direction of travel.

\

# Pitch Angle

As the AOA increases, the air circulating on the aerodynamic profile is diverted over a greater distance. That leads to an increase in airspeed and an increase in the lift force. As the AOA increased, it's difficult for air to flow smoothly over the top of the airfoil. At this point, the airflow will begin to separate from the airfoil and enter a turbulent pattern.

Turbulence could cause a sharp increase in the drag force and a loss of life in the area where it occurs. Increasing the AOA will increase the lift until the critical angle of attack is reached. Any increase in the AOA beyond this point will result in a failure and a rapid decrease in resilience. Various other factors could change the blade of the AOA rotor.

You can control the AOA only via the flight controls. Collective and cyclical springing will help to make these changes.

Helicopter pilots must use the cyclical springs to control the rotor system's position and control the rotor's backward tilt. To keep a hover at a constant height, the lift must correspond to the helicopter's weight. The pressure must equal to the pressure exerted by the wind and the tail rotor to maintain this position.

# The tendency to drift

During the hover, a single main rotor helicopter tends to move towards the tail rotor thrust. This lateral or lateral movement called a tendency to drift.

The helicopter's fuselage is also tilted when the tail rotor under the main rotor disc increases torque. The fuselage pitch is caused by the incomplete balance of the tail rotor thrust against the same plane's main rotor torque.

The helicopter tilts due to two separate forces, the main rotor disc tilting to neutralize the translational tendency and the lower tail rotor pushing below the rotating force effect level.

The vertical fin or stabilizer is often designed with the tail rotor attached to correct this lateral slip and eliminate certain tilt movements during hovering on some larger helicopters.

# The Coriolis forces

The rotation of Earth created a force known as the Coriolis force. Coriolis force deflects the airflow to the right causing winds above the friction level to flow parallel to the isobars.

That causes winds to flow clockwise around high-pressure areas and counterclockwise around low-pressure areas. In the southern hemisphere, the Coriolis force causes a counterclockwise flow around highs and a clockwise flow around lows.

The amount of deflection the air makes relates directly to the speed at which the air moving and its latitude. Therefore, slowly blowing winds will deflect only a small amount, while stronger winds will be more deflected. Likewise, winds closer to the poles will deflect more than winds at the same speed closer to the equator. A parcel of air, which resists convection when forced upward, is called stable air. Stable air could force uphill to form a mountain wave. Rising air, warmer than the surrounding air, describes as unstable air. A station circle used for the weather stations send up radio sounds — a satellite identified by a star on the map.

# Engine Anti-Ice

The anti-icing system found on most turbine-powered helicopters uses engine bleed air. Bleed air in turbine engines compressed air taken from within the engine, after the compressor stage(s) and before the fuel injected in the burners. The bleed air flows through the inlet guide vanes and the inlet itself to prevent ice formation on the hollow vanes. A pilot-controlled, electrically operated valve on the compressor controls the airflow. Engine anti-ice systems should on before entry into icing conditions and remain on until exiting those conditions. Use of the engine anti-ice, the system should always be following the proper RFM.

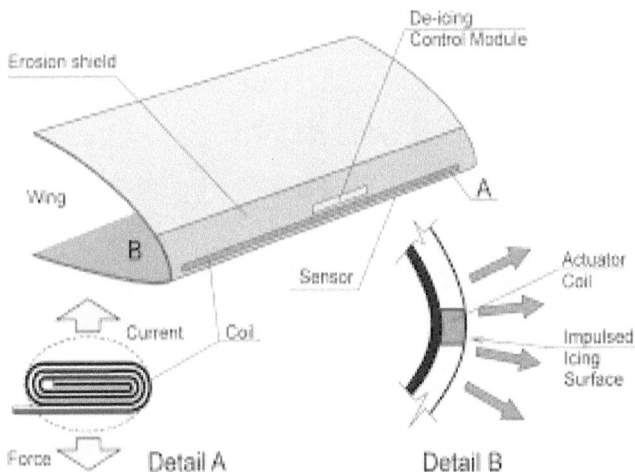

# Airframe Anti-Ice

Airframe and rotor anti-icing may find on some larger helicopters, but it not common due to the complexity, expense, and weight of such systems. The leading edges of rotors may heat with bleed air or electrical elements to prevent ice formation. Balance and control problems might arise if ice is allowed to form unevenly on the blades. Researches being done on lightweight ice-phobic (anti-icing) materials or coatings. These materials placed in strategic areas could significantly reduce ice formation and improve performance. The pitot tube on a helicopter very susceptible to ice and moisture buildup as well. To prevent this, they are usually equipped with a heating system that uses an electrical element to heat the tube.

# Deicing

Deicing the process of removing frozen contaminants, snow, ice, and/or slush from a surface. Deicing of the helicopter fuselage and rotor blades critical before starting. Helicopters unsheltered by hangars subject to frost, snow, freezing drizzle, and freezing rain can cause icing of rotor blades and fuselages, rendering them unflyable until cleaned. Asymmetrical shedding of ice from the blades can lead to component failure, and shedding ice can dangerous as it may hit any structures or people around the helicopter. The tail rotor is very vulnerable to shedding ice damage. Thorough preflight checks should be made before starting the rotor blades. If any ice was removed before starting, ensure that the flight controls move freely. For those helicopters with them in flight, deicing systems should be activated immediately after entry into an icing condition.

# Anti-Ice

Due to the complexity, expense, and weight of Airframe and rotor anti-icing systems on some larger helicopters, it not common to find. To prevent ice formation on the rotors' leading edges, we need to heat it with the bleed air or electrical elements. If ice is allowed to form unevenly on the blades, it might cause problems with the helicopter's balance and control. The Pitot tube on a helicopter very susceptible to ice and moisture and usually equipped with a heating electrical element system to heat the tube. Deicing the process of removing frozen contaminants, snow, ice, and/or slush from a surface. Deicing of the helicopter fuselage and rotor blades critical before starting. Helicopters unsheltered by hangars subject to frost, freezing drizzle, and freezing rain can cause icing on the rotor blades and fuselages, rendering them un-flyable until cleaned. Unequal peeling of ice from the blades can lead to component failure, and ice may hit any structures or people around the helicopter. The tail rotor is very vulnerable to shedding ice damage.

Thorough preflight checks should be made before starting the rotor blades, and if any ice was removed before starting, ensure that the flight controls move freely. While in-flight, helicopters equipped with deicing systems should be activated immediately after entry into an icing condition.

# Engine Anti-Ice

turbine-powered helicopters use engine bleed air for the anti-icing system. Bleed air is a compressed air taken from within the engine, after the compression stages and before fuel injected in the burners. The bleed air flows through the inlet vanes to the inlet to prevent the ice formation on the hollow vanes. The helicopter pilot-control the air flows with by switching an electrically operated valve on the compressor.

Engine anti-ice systems should on before entering into icing conditions and remain on until exiting those conditions. The use of the engine anti-ice system should follow the proper RFM.

When you using reduced power, such as during a descent, carburetor icing can occur during any phase of flight, and it's dangerous. You may not notice it during the descent until you try to add power.

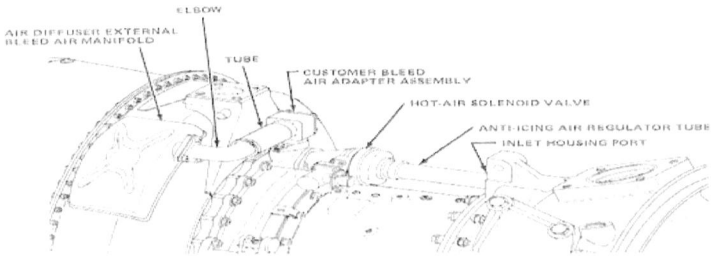

The carburetor air temperature gauge indicating a temperature outside the safe operating range, engine roughness, decrease manifold pressure, or decrease in engine speed, all indications of carburetor icing.

Changes in rpm or manifold pressure can occur for several reasons. The carburetor air-temperature reading on the gauge should check when carburetor icing conditions are necessary—the carburetor air temperature gauges marked with a yellow caution arc or green operating arcs. It is best to keep the needle out of the yellow arc or within the green arc in most cases. The carburetor heating system eliminates the ice by rerouting the air through a heat source before reaching the carburetor across.

# Cold Weather Operations

The aircraft always exposed to accumulations of frost, ice, or snow; the contamination should remove following the aircraft manufacturer's recommended procedures methods before a flight. The aircraft should pre-heat, when possible, before attempting to activate any of the aircraft systems. Cabin systems must properly handle to prevent damage from freezing or during defrosting.

Airplanes parked outside should fly against the wind as much as possible to minimize the accumulation of frozen precipitation in the openings around the flight control surfaces. Lids and plugs must install to protect the engine inlets from accumulation while the aircraft parked. The parking brakes should release after the aircraft brakes to avoid damage from temperature changes.

# Frost, Snow, and Icing

A takeoff should not attempt if the aircraft has frost, snow, or ice adhering to the windshield, or the instrument system, wings, rotors, control surfaces, or any other areas, which affect flight characteristics.

If frost, snow, or ice accumulations found on the Aircraft, it must de-ice before the flight by using industry-standard procedures and materials. If an accumulation continues to reoccur, you should make sure that the aircraft has been de-iced as often as necessary based on product specifications.

No flight should attempt into known or forecasted icing conditions.

# HELICOPTER
# OPERATION

# Standard Procedure

To ensure an increased level of awareness, the following procedures are the standard:

Take-off and landings will be made from adjacent taxiways or runways and not from the ramp.

There will be no flight of other aircraft at low altitudes over or near the helicopter.

Helicopters will avoid hovering on the ramp or near other aircraft.

When ground taxiing, you should reduce power to minimize rotor wash, and verbalize any safety concerns before taxiing.

Overlapping of the rotor blades and airplane wings will be avoided

When parking, the helicopter's tail should move away from the passengers' direction.

At the helicopter controls, you should be attentive to ground personnel's movement by staying on the controls and ensuring that the rotor disc tilted aft far enough for passenger clearance.

When near the running helicopter, Hats should remove. No umbrellas will be used near or under the helicopter rotor disc.

At no time running is permitted around the helicopter.

Constant vigilance shall maintain by the crews when limos or autos operating on the ramp, never allow them to drive under the rotor disc.

# Helicopter Rotor Brake

The rotor brake on the helicopter is not used when the aircraft engine starts. However, if the wind is too strong or in other unusual circumstances, the rotor brake can be used during starting, with the engine started before the rotor brake is released.

All individuals must stay away from the rotor with the rotor brake applied when starting the engine. Rotor brakes are known to slide, which requires immediate action by you to fully release the brake or shut down the engine.

Rotor brakes are fitted to slow the rotors down. A simple tie-down is all that is required to keep the rotors still when not moving. The Blades rely heavily on Centrifugal force to keep them up and away from the tail boom and above passengers' heads when under the disk. As the blades slow down, there is very little control lift to keep them where you want them, so the idea is to take them from the lower control speed to stationary as quickly as possible.

All rotor brakes are essentially similar in design to a car disc brake, with 2 or 4 pistons activating calipers on either side of the brake disc. Robinson is the only non-hydraulic system used. All others are usually via a large over-center hydraulic piston in the cockpit. Some helicopters (such as the Sea King) can apply the rotor brake with the engines still running at idle. Some have an aux drive that bypasses the MGB and allows blade folding: the rotor brake is one of the interlock requirements.

# Baggage

You should position the baggage in the special compartment to ensuring they do not interfere with an emergency by blocking access to the aisle. During cruise flight, the passengers may access their hand baggage but must secure it for takeoff and landing. You must brief the passengers of this requirement.

Loading passengers' personal belongings the responsibility of the flight crews and ground personnel.

# Basic Flight Maneuvers

As soon as a helicopter leaves the ground, it is exposed to four aerodynamic forces. Thrust, pull, lift, and weight. Understanding how these forces work and how they controlled with electrical and flight controls essential for flight.

They defined as follows:

• Thrust is a forward force that is generated by the rotor. It overcomes resistance.

• Drag is a rearward retarding force caused by the disruption of airflow by the rotor, fuselage, and other extended objects. Drag force opposes the thrust and acts backward and parallels to the relative wind.

• Weight is the helicopter's total combined load, including the crew, the fuel, and cargo or baggage. Because of the force of gravity, the weight pulls the aircraft downward. It opposes the Lift and acts vertically downward through the helicopter's center of gravity (CG).

• Lift is opposing the downward force of weight and is produced by the dynamic effect of air acting on the airfoil. Lift acts vertically to the flight-path through the center of lift.

Because of wind's variation and density altitude, two flights rarely alike to pilots even if they fly at the same altitude, hence, it essentially impossible to prescribe helicopter altitudes for the performance of each flight maneuver.

Altitudes, airspeeds, and power settings will constantly vary due to the loading, weather, and the specific helicopter.

The sterile cockpit environment should also be maintained during crucial flight periods, i.e., Approaches and Departures. Each flight crewmember has a headset and must use it anytime the aircraft in motion on the surface or airborne.

The aircrew should also eliminate all inessential cockpit conversation that does not apply directly to the aircraft's operation anytime the aircraft is in motion, particularly while operating in congested, busy airspace.

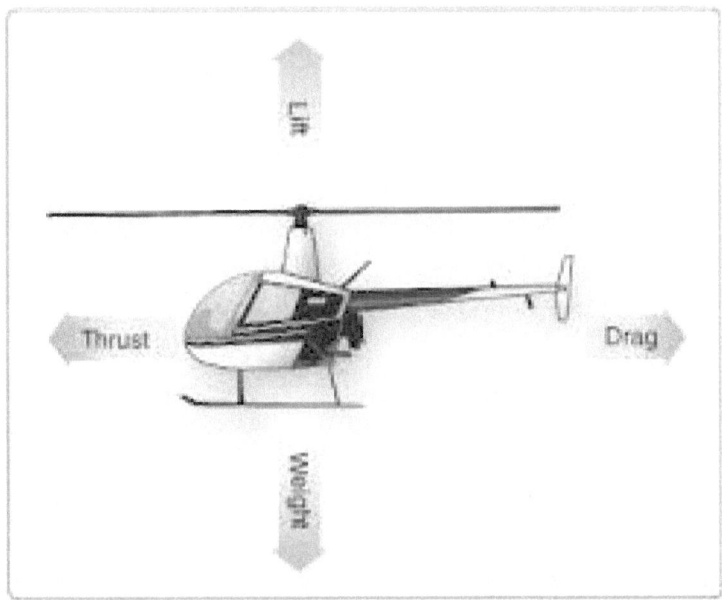

# Boarding

Whenever the helicopter engine is operating, at least one pilot must be at the flight controls. You need to be watchful to personnel and equipment near the aircraft. When boarding or deplaning the helicopter, a crewmember shall always escort the passengers. The passengers should never approach or depart the aircraft from the front due to the main rotor blades' low-tip path. Passengers should never allow moving beyond the baggage compartment area or toward the tail rotor. At least one crewmember must supervise the passenger movement and ensure it's done from the same side of the helicopter.

At least one crewmember will load or unload the passengers' baggage and confirm that the passengers are seated with the seat belts fastened. Each of the passengers will be handed a briefing card.

One of the crewmembers should confirm that all doors are properly secured before returning to the cockpit.

# Hovering

A helicopter hovers when it maintains a constant position over a point on the ground, usually a few feet above the ground. To hover, a helicopter's main rotor must supply lift equal to the helicopter's total weight, including crew, fuel, and, if applicable, passengers, cargo, and armaments.

The rotor system requires a large volume of air when hovering. This air is pulled from the surrounding air mass; this is an expensive maneuver and requires a great deal of engine power. The air is supplied by the rotary blades and drawn in from the top at a relatively high speed, forcing the rotor system to operate in a descending column of the air environment. Hovering is one of the most important maneuvers and complicated matters. A change in speed or vehicle speed readily affects the performance required to maintain the altitude. This affects the rotational force. You, as a pilot, need to make changes and fix them quickly.

The key operation is to focus on objects about 30 to 50 feet away, or at the horizon, so that you will able to sense the movement, stop it, and immediately return to the original position. Many helicopter pilots have a habit of looking directly down at the skids to determine when they'll touch the ground. It should not be done this way.

You should look for small changes in the helicopter's altitude and attitude. When these changes are noted, you should make the necessary control inputs before the helicopter starts moving away from that point. In normal hovering, the helicopter's skids should not be more than three to five feet off the ground.

From this midair position, you will eventually use pedals turning and hover-taxi the aircraft. To set the helicopter down, you should gently reverse the same inputs used to lift it into the air. As the main rotor is now producing less torque, it's essential to gently lower the Collective and reduce the left pedal pressure.

As with a takeoff, you need to control the Collective's altitude to maintain a constant RPM with the throttle. The pedals are used to control the heading while the cyclic maintain the helicopter's position. You need to make only small and coordinated corrections to maintain the helicopter in a stabilized hover. When that's achieved, you can remove the correction to stop the helicopter's movement. A subtle sense for the helicopter movement must be developed by you because hovering too low will result in an infrequent touchdown. Dynamic rollover accidents usually occur over a level surface.

# Hovering in Flight

The most challenging part of flying a helicopter is hovering. The reason being is that all helicopters are generating their own gusty air while in a hover. This wind is acting against the flight control surfaces and the fuselage. It forces you to keep the helicopter in position despite the inconvenience.

You can use the cyclic to eliminate drift in the horizontal plane while controlling the forward, backward, right, and left movement. To control engine speed, you must use the throttle if there no governor control. You need to use the collective to maintain altitude while controlling the heading or nose direction with the pedals. What makes hovering so difficult is the interaction between these controls. An adjustment in any one control requires an adjustment of the other two, making a cycle of constant correction necessary. A helicopter could maintain a constant position over a selected point during hovering flight. Usually, its a few feet above the ground. This ability of the helicopter to hover comes from both the lift factor and weight.

Thrust is created by the rotor system. It's directed by the pilot that uses the Cyclic to compensate for the wind and hold a position. When hovering in a no-wind condition, all the opposing forces (lift, thrust, drag force, and weight) balance. Thus, when lift and weight are equal, the helicopter will remain at a stationary hover. The amount of the main rotor's thrust can be changed while hovering to maintain the desired altitude.

It's done by moving the Collective to change the angle of attack (AOA) of the main rotor blades. When changing the AOA, the drag force on the rotor blades change, and the engine's power change as well, so to keep a constant rotor speed.

If the lift is greater than the actual weight, the helicopter will accelerate upwards until the lift force is equal to the weight gain altitude. If the thrust is less than the weight, the helicopter will speed downward. When hovering is just above the ground, the proximity to the surface will change this response. While the rotor blades producing the lift, the hovering helicopter drag force becomes an induced drag force.

A tail rotor produces thrust in the opposite direction to the rotating force, but the tail rotor's thrust is sufficient to move the helicopter sideways.

# Hovering turn

A hovering turns a maneuver, which performs at hovering altitude when the helicopter's nose rotates either left or right while maintaining the same position over a reference point on the surface. This maneuver requires coordination between all flight controls, and precise action, especially near the surface.

Hovering turns are performed around the mast or the tail of the aircraft. At a constant altitude, a rate of turn, and engine speed should all be maintained by the helicopter pilot, who can initiate the turn in either direction by applying the anti-rotating force pedal pressure toward the desired direction.

More power is required during a left turn because the left pedal will increase the tail rotor's pitch angle and require more engine power. Less power is required when turning to the right.

To maintain the helicopter on the selected surface reference point, a hovering turn should be avoided when the winds are strong enough to preclude sufficient aft cyclic control.

# A go-around a procedure

A go-around procedure is done when the helicopter remain airborne after the intended landing is discontinued. A go-around may be necessary when:

• Instructed by the control tower.

• Traffic conflict occurs.

• The helicopter is in a position from which it is not safe to continue the approach.

Any time an approach uncomfortable, incorrect, or potentially dangerous, abandon the approach. The decision to make a go-around should be positive and initiated before a critical situation develops. When the decision is made, carry it out without hesitation. In most cases, when initiating the go-around, the power is at a low setting. Therefore, the first response is to increase the collective to take-off power. This movement is coordinated with the throttle to maintain engine speed and proper antitorque pedal to control your heading. Then, establish a climb attitude and maintain a climb speed to go-around for another approach.

During a crosswind approach, crab into the wind. At approximately 50-100 feet of altitude, use a slip to align the fuselage with the ground track.

The rotor is tilted into the wind with cyclic pressure so that the helicopter's side movement and the wind drift counteract each other.

Maintain the heading and ground track with the anti-torque pedals. Under crosswind approaches, the ground track is always controlled by the cyclic movement.

The heading of the helicopter in hovering maneuvers is always controlled by the pedals. The collective controls power, which the altitude at a hover. This technique should use on any type of crosswind approach, whether it a shallow, normal, or steep approach.

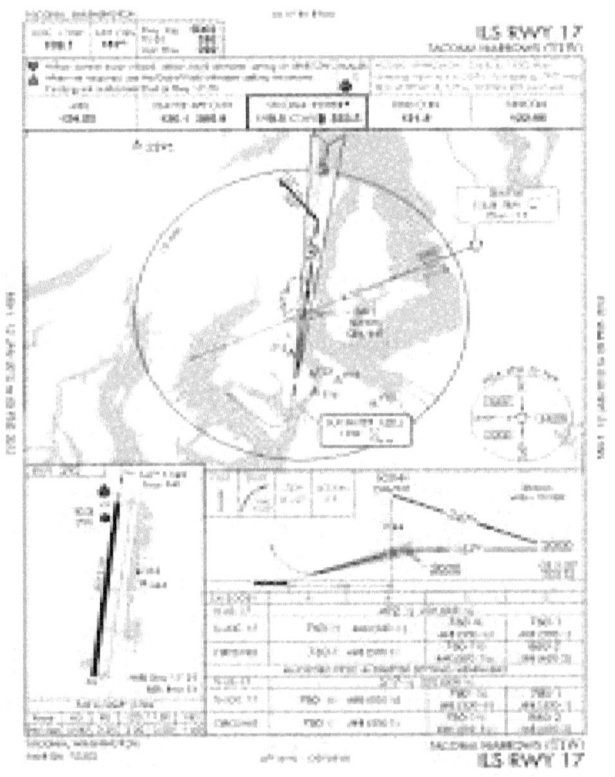

# Taxiing/air taxi

Air taxi the preferred method for helicopter movements at airports provided ground operations permit. Air taxi authorizes you to precede above the surface via either hover-taxi or flight at speeds more than 20 knots. You are expected to remain below 100 feet AGL. You are solely responsible for selecting a safe airspeed for altitude and operation. Issuing takeoff clearances from movement areas other than the active runway instead of extended hover-taxi or air taxi operations.

If takeoff is requested from the area not authorized for helicopter use or outside the airport, you need to use the correct phraseology. When necessary for a wheeled helicopter to taxi on the surface, it uses less fuel than hover taxiing to minimize air turbulence. Under rough, soft, or uneven terrain, it's necessary to hover/air-taxi for safety considerations.

Helicopters with three or more main rotor blades are subject to ground effect and may suddenly lift off the ground. You must proceed in the ground effect above the surface at a slow speed when requesting vertical take-off or landing.

When taxiing from one area to another at the airport, you should normally use this technique to move the helicopter a short distance. Taxiing in a helicopter is often known as hover- taxiing because you will hover just a few feet off the ground with a forward motion.

If you lift the skids more than three feet above ground level, the helicopter will effectively fly out of ground effect and need about 10 percent more power to maintain a hover. The helicopter may not hover out of ground effect under certain conditions, which could have some vegetation, steep or rough terrain, or high altitudes.

When taxiing, you, the pilot, must sit in the pilot seat, with seatbelts fastened. You are not allowed to perform any other cockpit duties while taxiing close to other aircraft or obstructions. You cannot taxi the helicopter unless there is enough clearance to maneuver safely. You should utilize the ground attendants to maintain the required clearance. The helicopter should be towed whenever the clearance is insufficient.

Airplane wheels are free turning. They have brakes, but no engines or motors to drive the wheels. Airplanes move forward on the ground the same way they do in the air.

Some jets will save fuel and taxi on one engine, then start the other shortly before pulling onto the runway to takeoff.

The helicopter should be stationary on the surface with the collective is full down. Move the cyclic slightly forward and apply gradual upward pressure on the collective to move the helicopter forward along the surface.

Use the anti-torque pedals to maintain the heading and the cyclic to maintain ground track. The collective controls starting, stopping, and speed while taxiing. The higher the collective pitch, the faster the taxi speed; however, do not taxi faster than you walk. If the helicopter is equipped with brakes, use them to slow down. Do not use the cyclic to control ground speed.

During a crosswind taxi, hold the cyclic into the wind a sufficient amount to eliminate drifting.

Helicopters usually hover with the left side low due to the tail rotor thrust being counteracted by the main rotor tilt when the nose in a low or high condition.

Taxiing accrues on or near the surface or other routes.

There are three types of taxiing. A hover taxi used when flying below 25 feet above ground level. An air taxi is preferred in greater distances within an airport boundary. The helicopter should fly below 100 feet and avoid flying over vehicles or people.

# Takeoff and Landing

This system approach was derived from all aircraft, whether fixed wing or helicopter, once airborne, must have a place to land. Fixed-wing aircraft require large airports with long runways, taxiways, aprons, control towers, and hangars. All airport lighting systems are integrated and involve hundreds of different lighting fixtures. Each type is being important to the whole system. Helicopters increasingly have to land at night, and they also need an integrated system, but their landing requirements are different. When landing at slow speed and a steep angle of descent, stop and hover before the landing. You must see the ground with good depth perception to land safely. This means that helicopters are requiring an entirely different and properly designed heliport lighting system.

The Downing Heliport Lighting System is a comprehensive, coordinated, and innovative lighting system designed especially with the helicopter operator in mind.

All of the visual aids in the Downing Heliport Lighting System work independently and together to make landing a helicopter at night as safe as landing it in the daylight. All of your heliport needs – from one single source. It has a set of Portable Heliport Lights, complete with Surface Floodlights, Perimeter Lights, a VASI, and a Locating Beacon packaged in a steel dolly type of portable carrying rack.

It contains all of the connecting electrical wiring, connectors, and junction boxes needed to complete one electrical circuit for the system.

It's powered by a small gasoline generator, and it assembled in about 1/2 hour. It weighs about 250 pounds and can be transported either by a sling or inside the helicopter.

The (TLOF) area is a configuration that provides a greater illumination and better uniformity to the touchdown and Liftoff (TLOF) boundaries. Model 700 Series have an improved design to blocks the flood lamps' visibility as seen from two or more feet above the TLOF surface or within the TLOF area. This makes the night time access and outlet to and from the helicopter much more comfortable.

# Takeoff/landing regulations

One of the big advantages of helicopters is the ability to land off-airport. Deciding where and when to land a helicopter worthy of studying as the consequences of a bad decision can very serious.

The FAA Regulations title14 CFR Part 91.13 prohibiting careless or reckless operation but not addressing the landing on someone else's property. It could construe as trespassing if you did not obtain permission from the landowner before landing.

The zoning laws prohibit the landing of aircraft even with the owner's consent. Some cities have specific ordnances, which require a landing permit, and if one are not obtained even if proper security measures are taken, it can result in a heavy fine. Besides, trained ground personnel (police, fire personnel, and so on...) must secure the area of landing in the unsecured area and prevent unauthorized individuals from potentially approaching the helicopter when it unsafe.

Criminal charges can file if somebody or properties placed at risk during a landing. Under part 91.13, the FAA can attempt to charge you with this violation.

A helicopter pilot who is not designated as a High Minimums Captain may take off from a runway providing that:

- ➢ That runway at or above 800 feet and the visibility 1/8 statute mile,
- ➢ A suitable takeoff alternate exists within 75 NM.

> The helicopter pilots shall determine before takeoff from weather reports, forecasts, and NOTAM's, that the alternate at or above landing minimums and expected to remain so for the time required.

> Helicopter pilots must follow and obey the published noise reduction procedures and curfews except when the flight's safety is involved. An optimal location for an off-Airport/Heliport landing site close to the location of the passengers.

> The helicopter pilots should use their best judgment when considering a potential landing site.

A suitable landing site must a minimum of 75 x 75 ft. The location may differ. It could on the ground, an elevated platform, a marine vessel, or the roof of a building. A suitable landing area will provide a visual means to determine the wind direction. The landing area may not contain loose objects, blowing up into the rotor system or the engine intakes.

Flight Dispatcher or the helicopter pilots must confirm with the local authorities that the helicopter landing authorized at the location before landing at an Off-Airport site. An inspection of the site is required before issuing landing permission.

The landing surface must be dry, firm, leveled, and able to support the aircraft's weight and have at least one clear approach and departure path.

When the helicopter operates on the ground, the passengers must be provided safety at all times.

The Passengers must never approach or depart the helicopter unless they are escorted by crewmember or ground personnel who trained in helicopter safety procedures.

The responsibility for determining that this requirement met is in the pilot's handbook.

# Approaches

Helicopter approaches have more to do with local traffic and terrain than with target speed and configuration. Enter the airport traffic area safely and avoid obstacles. Follow the landing procedures as described.

One approach is a transition from the height of the traffic pattern to a hover or surface flight. The approach must end at hover altitude and, at the same time, reach the rate of fall and zero ground speed.

You should always approach a specific and previously defined landing site. A normal approach uses a descent profile between 7 ° and 12 °, starting at around 300-500 feet AGL.

On the final approach, the helicopter should be at the recommended approach speed and at an AGL of approximately 300 feet on the correct ground lane (or ground orientation) for the intended landing site. Still, the Helicopter centerline must first align at approximately 100 'AGL for control. Shortly before reaching the desired angle of approach, lower the collective so that the helicopter will slow down and lower the approach's angle.

As the collective decreases, the nose will lean back, so the back cyclic must maintain the recommended approach speed level. Adjust the anti-rotation power pedals as needed to maintain the trim.

The most important for a normal approach is to maintain a constant approach angle at the endpoint. Use the Cyclic to control the closing speed or helicopter speed to the point of touchdown.

\* \* \* \* \*

The landing approach is the same as the normal approach to a hover; however, instead of terminating at a hover, continue the touchdown approach. Touchdown should occur with the skids level, zero ground speed, and a rate of descent approaching zero.

As the helicopter comes near the surface, you increase the Collective, as necessary, to cushion the landing on the surface, and terminate a skids-level altitude without straight forward movement. You should also consider forced landing areas in case of an emergency. A high reconnaissance should fly at an altitude of 300 to 500 feet above the surface. As a general rule, ensure that sufficient altitude is available at all times to land into the wind in case of engine failure. Always maintain safe altitudes and airspeeds, and keep a forced landing area within reach whenever possible.

\* \* \* \* \*

The helicopter takeoff procedures can begin from a runway, taxiway, a portion of a runway, or any clear area, which could use as a landing area such as a rooftop of building, ramp, heliport, and helipad.

Whenever possible, the takeoff clearance will be issued in connection with a hover or air taxi operations. Please look at the Phraseology in the appropriate chapter.

Wind indications near the helicopter should be alerted to the Helicopter pilots. The intended method of departing should be advised to ATC. You must give a direction that is indicated that you are willing to accept the wind condition and controllers' advice.

The Departure points could significantly distance from the control tower but could also be impossible for the controller to determine the helicopter's relative position to the wind.

This procedure allows you to lift off into a hover a few feet above the ground and then accelerate to 60 knots before initiating a slow climb.

The wind direction and speed should be noted first. You should then take off directly into the wind to minimize the sideway drift and increase the helicopter's performance during a climb.

When facing a fifteen-knot wind, the rotor will experience a translational lift even on the surface.

When you ready to lift off, you must use scenery objects as a guide and note a point in the distance. That point and the outside horizon should be used as references to maintain the helicopter's alignment and altitude as you lift-off.

The Cyclic should be set in and around the neutral position, and the Collective should in a complete down position.

To continue the liftoff, you should smoothly increase the Collective, anticipate the need to add left pedal as you lift off, and then make small, smooth corrections with the Cyclic and pedals.

You should hold the helicopter skids about 3 feet (1 meter) above ground, staying low in case of engine failure and keep the helicopter in the ground effect.

At this time, you should raise or lower the Collective to maintain the correct altitude using small cyclic pressures, and the force pedals to keep the helicopter's nose from rotating.

A slight forward cyclic pressure is needed when taking off into a headwind, a left pressure with a left crosswind, etc. you must apply a gentle straight forward cyclic force to lower the nose and begin moving forward along the departure path. The helicopter may tend to settle as you move forward. To avoid this, you can compensate by adding a slight upward Collective.

As the helicopter airspeed reaches 10 to 15 knots, it enters the effective translational lift phase.

The nose will tend to yaw left and slightly pitch up. Some Cyclic forward needs to be applied to prevent the nose from rising.

As the helicopter continues climbing and accelerating, the takeoff is continued by flying in a different pattern, climbing to 300 feet at 60 knots, while maintaining a near nose-level altitude. Then make 90 degrees turn to the left or right toward the crosswind leg as you climb to 500 feet and maintain 60 knots airspeed.

\* \* \* \* \*

For landing, a slope of 5° is considered the maximum for most large helicopters' normal operation. As the collective is lowered, you should continue to move the cyclic toward the slope to maintain a fixed position. The slope must be shallow enough to hold the helicopter against it with the cyclic during the entire landing.

The ideal landing zone is a level, 100-by-100-foot, or larger area of grass or hard surface. Most helicopters with a main rotor diameter of 35-50 feet and a fuselage length with main rotor blades turning 40-50 feet.

If the takeoff requested from an area that is not authorized for helicopter use, and is not visible from the tower, or unlighted area at night, the ATC will inform you 'AT YOUR OWN RISK' and may provide additional instruction. Ultimately, you are responsible for safely operating the helicopter, and you should avoid unnecessary risks.

Every effort made to permit helicopters to proceed directly and land as close as possible to their final destination at the airport.

Helicopters need more taxiing instructions that may affect how the service can be expedited.

As with any ground movement operations, your cooperation highly necessary to achieve safe and efficient takeoff maneuvers.

ILS glide path coverage in elevation. Glide path signals available from 0.45 $\theta$ to 1.75 $\theta$ concerning the horizontal passing from the touchdown, where $\theta$ the glide angle.

Visual ground aids, such as VASI (visual approach slope indicator), which provide vertical guidance for a VFR (visual flight rules), approach the visual portion of an instrument approach and landing.

PAR (precision approach radar), which is used by the ATC (air traffic control) to inform an aircraft making a PAR approach of its vertical position (elevation) relative to the descent profile.

These areas are separated or located on an airport/heliport, and ATC will issue takeoff clearances from an area other than an active runway, with additional instructions.

* * * * *

A running take-off is designed for helicopters with fixed landing gear, such as skids, skis, floats, or wheels. It's used when the conditions of load and density-altitude prevent a sustained hover at normal altitude. A rolling takeoff is occasionally used for wheeled helicopters to minimize the downwash created during a hover takeoff. A running maneuver should be avoided if the helicopter has not achieved the necessary power to hover.

# Vertical Takeoff

Helicopter pilots do not normally take off in the sky or fly quickly close to the ground unless there are obstacles to avoid. This maneuver is requiring a high level of concentration and coordination.

The rotor blades of the helicopter will create a vertical thrust. During a vertical takeoff in hover, the helicopter flies perpendicular to the ground while maintaining a constant heading up to a skid height of two to three feet.

If the helicopter hovers a few hundred feet above the ground at low speed or no speed or moves very quickly near the ground, you will have difficulty recovering in the event of an engine failure. Damage to the engine at low altitude at low speed can increase the risk to the helicopter. Without making a safe landing with an automatic landing, even a bug cannot fix it.

# Maximum Performance Takeoff

A maximum power takeoff is used to climb at a steep angle to remove obstacles on the flight path. Vertical takeoff is recommended but not preferred when taking off in areas surrounded by high obstacles.

Out of Ground Effect (OGE) is where there are no hard surfaces for the downwash to react against. For example, a helicopter hovering 160ft above the surface will be in an OGE condition and require more power to maintain a constant altitude than hovering at 15ft. Therefore, a helicopter will always have a lower OGE ceiling than in ground effect (IGE) due to the available engine power.

You need to familiarize yourself with the equipment's features and limitations before attempting to make maximum performance. To safely carry out this type of take-off, sufficient power must be available to move the OGE so that the helicopter cannot land again after take-off. A gravity check can be used to determine if there enough force to perform this maneuver. For a start with maximum performance, the climb angle depends on the existing conditions. Usually, this maneuver started from the surface.

Before taking off at maximum power, you must position the helicopter in the wind and bring it to the surface to allow a longer take-off. The helicopter must then be covered. You must also determine the excess power available by defining the difference between the available power and the power required to hover.

After checking the area for obstacles and other aircraft, you must select reference points along the take-off path and keep the path on the ground.

The rotor's maximum speed must be maintained to not decrease since you would probably have to lower the collective to regain control. These entrances have to wait until the helicopter has overcome the obstacles.

# Takeoff from a Hover

When you already hover in the air, the transition to a forward flight is executed by increasing altitude quickly. Before initiating the take-off, ensure that you have completed the proper checklist, and all helicopter systems are within limits.

After systems check, bring the helicopter to a hover. The power check should include the power available; that is, the difference between the power being used to hover and the power available at the prevailing attitude. Visually clear the area. Start moving the helicopter by slowly easing the cyclic forward. As the helicopter starts to move forward, increase the collective to prevent the helicopter from sinking and adjust the throttle to maintain engine speed.

This increase in power will require an increase in the antitorque pedal to maintain your heading. Keep a straight takeoff path throughout the process while accelerating as the helicopter starts climbing, and the nose tends to rise due to the increased lift force. Now you can adjust the collective to obtain a normal climb and apply enough forward cyclic to overcome the nose's tendency to rise. Hold an attitude that allows you to accelerate airspeed and gain altitude to complete the takeoff without complications. Maintain a more favorable climb formation.

As the helicopter continues to climb and accelerate, apply aft cyclic pressure to raise the nose to the normal altitude.

Failing to use sufficient collective pitch to prevent loss of altitude and adding power too quickly at the beginning of the take-off, without a straight forward cyclic compensation, will cause the helicopter to gain altitude before acquiring a proper airspeed, a mistake you should avoid.

# Rearward Flight

As a pilot, you may be required to move the helicopter to a specific area where a forward or sideward hovering flight cannot make.

The area behind the helicopter must be cleared before beginning this maneuver. The limited visibility suggests that you use ground attendants. You find out two reference points in front, in line with the helicopter, just like you do when hovering forward. Apply a rearward pressure on the cyclic, and after the movement has begun, position the cyclic to maintain a groundspeed no faster than a walk. Maintain constant ground track with the cyclic, constant heading with the anti-torque pedals, and constant altitude with the collective. Keep your engine speed steady.

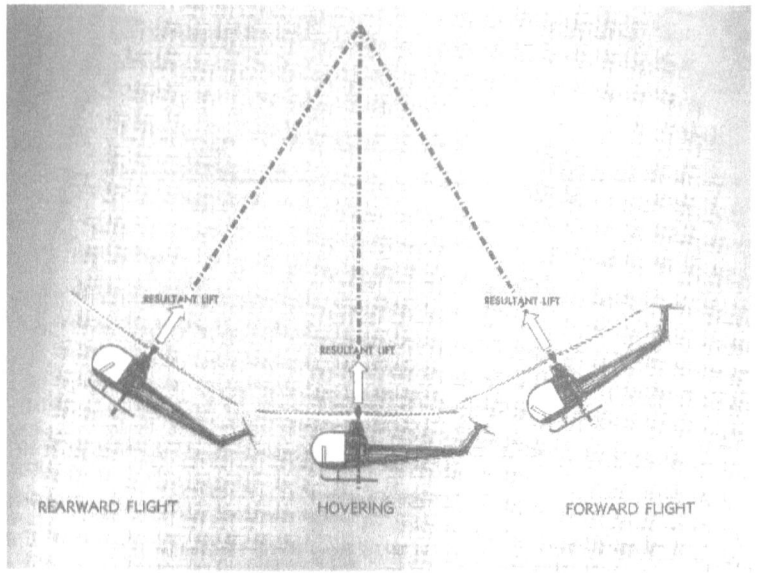

# Flight Log-checklists

Logbooks matter more than you think - here's what you need. Logbooks aren't just a way to keep track of your flight times and training records. They're a representation of you as a pilot, your personality, and your skill sets. If you're applying for a job in the cockpit, you're being hired not for the certificate you hold, but for the experiences and knowledge you've built over time that will make you a successful and safe professional pilot. In this process, logbooks might be more important than you think. In some cases, they'll determine your ability to get hired.

The first logbooks are usually the messiest and disorganized. It makes sense; as you start flight training, you're bound to make a few mistakes, and that's ok. But if your goal is to be a professional pilot, you should try log flights as a professional from day one. Does this mean you need to be perfect? Nope. It just means it's something you should care about and spend effort on each and every time you write an entry.

A logbook is essentially a legal document, so make sure it's legible, organized, signed, and totaled correctly. If you can, find a good electronic program to begin logging flight times alongside your paper copy early on. It will save countless hours later on.

Don't forget to document your logbooks and back them up with pictures or scans. Personally, I take a picture of every new totaled page and keep them in a file that's on my laptop and synced to an online server. If you lose the only copy of a logbook, you'll likely lose dozens of hours of progress, and much of it will be hard to re-record accurately. Protect them by keeping them somewhere safe.

We highly recommend that you use a reputable electronic program to log flight time. It's important to use a dedicated logbook application, instead of trying to format your own spreadsheet. Self-made spreadsheets contained the most errors, inaccurate information, and are generally formatted less professionally.

If you have 6 or 7 logbooks, that's ok. Just make sure each logbook is totaled at the end, and the times are carried forward accurately to the next logbook. Large professional pilot logbooks are the best, if possible.

In the airline world, pilots sometimes use small trip books to keep track of their flights and duty times before they have time to pull out the physical logbook for making entries. These trip books aren't professional documents and shouldn't be used as such.

When you do make an error, corrections are all about cleanliness. Cross through the mistake with a single line and correct it in the margin.

The most common logbook problems are unsigned pages, incorrect totals (especially when single and multi-engine times don't match up to total time), totals not matching between separate logbooks, and strange or incriminating comments written next to entries.

Make sure your electronic logbooks are correct and updated, logbooks are one great representation of you as a pilot.

An FAA review of helicopter incidents and accidents over the past five years has identified several accidents in which loss of control was encountered immediately after liftoff while light on the skids/gear, or from other issues caused by missed checklist items. With those operational issues in mind, the agency issued a safety alert to helicopter pilots about the use of checklists and a reminder to perform stabilized hover checks before takeoff.

Helicopter operations generally require a moment or two just after liftoff, to ensure that everything aboard is humming along smoothly before leaping into the air. Numerous loss-of-control accidents have been tied to factors pilots making a few precautionary checks following the engine start.

In one accident, the pilot attempted to lift off from a rooftop heliport without performing any kind of power check from the hover. As the helicopter moved over the rooftop's edge out of ground effect, it crashed to a parking lot below. The investigation showed that while one of the helicopter's two engines was accurately set to the "fly" position, the second had been left in the "idle" position, a fault that could have been identified during a hover check.

# Aircraft Maintenance Records

In-depth knowledge of the maintenance records and helicopter systems' operational requirements is a procedure and practice, showing compliance with general aviation maintenance.

Record-making and record-keeping requirements are under (14 CFR) parts 43 and 91.

The regulations state that aircraft records must contain a description of the work performed, the date it was completed, the signature of a certified mechanic's, and the certificate number of the person approving the aircraft for service return. The registered owner or operator must keep the following records of each aircraft's preventive maintenance for the last 100-hour of required inspections.

The records must include the signature and certificate number of the individual who approved the aircraft for service, including the status of applicable airworthiness directives (AD). The AD or safety directive involves any recurring action, time, and date of the next action required.

Maintenance records are important—their many important reasons for keeping good records.

Incomplete or missing records will greatly affect the value and the ability to sell an aircraft. The records should show the work performed, the time it was done, and who did the work.

The records provide written proof of the parts, labor, and any sublet services provided to the new owner with valuable information about the aircraft's history. The FAA provides an overview of the procedure's standards and dictates what information should document, who must record it, and when it must be documented in the maintenance record as a logbook entry.

The person doing the work needs to sign-off the completeness and accuracy of the aircraft's maintenance book. If an aircraft involved in an accident and someone was injured or killed, the investigator would want to see the maintenance records or inspection performed by a qualified person. Accurate maintenance archives can help show service history. All logs must be kept for the life of the aircraft. The regulations allow certain records to be discarded when the repair is repeated or superseded. Any reproduction or alteration of records or reports, for fraudulent purposes, a ground for suspension or revocation of the licenses or certificates.

# Guidelines practices

During the flight, those are good practices to follow:

When entering a maneuver and the trim, the rotor, or the rotating force, reacts quicker than is anticipated, and so you should move the Cyclic only as needed to keep up with the rotor. If you continued to move the cyclic, you have been exceeding your control, and the aircraft will reach its limit. You need to perform the maneuver with less intensity until all aspects of the helicopter move could be controlled. You should be aware of the flight control sensitivity due to the high speed of the main rotor.

You can predict changes in aircraft performance due to load or environmental conditions. The normal collective increase to check the rotor speed at sea level (SLS) may not be sufficient at 4000 feet (PA) and 95 ° F.

The cyclic position relative to the horizon can determine the direction and altitude of the helicopter.

# Confined area operation

A Confined Area is an area where the helicopter's flight is limited in some direction by terrain or obstructions, natural or manmade. For example, a clearing in the woods, a city street, a road, a building roof, etc., can each be regarded as a confined area. Generally, takeoffs and landings should be made into the wind to obtain maximum airspeed with minimum ground speed.

There are several things to consider when operating in confined areas. One of the most important is maintaining clearance between the rotors and obstacles forming the confined area.

The tail rotor deserves special consideration because, in some helicopters, you cannot always see it from the cockpit. This is not only applying while making the approach but while hovering as well.

Another consideration is that electric wires especially are difficult to see; however, their supporting devices, such as poles or towers, serve as an indication of their presence and approximate height. If any wind is present, you should also expect some turbulence.

Something else for you to consider is the availability of forced landing areas during a planned approach. You should think about the possibility of flying from one alternate landing area to another throughout the approach while avoiding unfavorable areas.

Always leave yourself a way out if the landing cannot be completed or a go-around is necessary.

A high reconnaissance should complete before initiating the confined area approach. Start the approach phase using the wind and speed to the best possible advantage.

Keep in mind areas suitable for forced landings. It may be necessary to choose between an approach with a crosswind, but over an open area, or one directly into the wind, but over heavily wooded or extremely rough terrain where a safe forced landing would be impossible. If these conditions exist, consider the possibility of making the initial phase of the approach crosswind over the open area and then turning into the wind for the final portion of the approach.

# Total Power in helicopters

Every helicopter's performance envelope is defined by the relationship between the power required and the power available at some flight conditions.

In other words, the helicopter does not care what kind of engine powers it, any more than the engine cares what it supplies power to.

It takes the same amount of force to hold a 5,000-pound helicopter one foot off the ground as it does to hold it at any cruising altitude. But to create that lift force, we need to turn the rotor and accelerate a mass of air. In the physics world, it said that the rotor does work on the air. It takes power (the rate at which work is done) to overcome the rotors' drag force and create lift. Power is a parameter that we can measure in the cockpit, and changing conditions of air density, airspeed, and flight conditions will all cause a change in the power requirement.

Generally speaking, the maximum power available from the engine can be assumed as constant. Circling back to the idea that power requires to be airframe dependent can further break down into the profile's subcategories, induced, and parasitic issues.

Profile power requires overcoming the friction drag force on the blades and pushing the rotor's shape through the viscous air. It does not change significantly with a change in the angle of attack and is accounts for 15% to 40% of the main rotor power required in a hover. It stays relatively constant with airspeed until high speeds, compressibility, and/or blade stall drive it up.

Induced power is the power required to overcome the drag force developed during the creation of a rotor thrust. With an increase in the attack angle, the airflow that moves down through the rotor causes the blade's total reaction lift vector to tilt rearwards, creating an induced drag force.

It takes around 60% to 85% of the total rotor power in a hover to overcome it. Parasitic power is the additional power required to move everything else attached to the rotor through the air — that's the fuselage and everything attached to it. It rises up with the square of the airspeed.

Adding up these three power components and miscellaneous power consumers like the tail rotor, hydraulic pumps, gearbox losses, generators, etc., results in the familiar total power required curve that defines our flight envelope.

Moving from hover into straight forward flight brings a rapid decrease in the required power due to a change in the oncoming air's inflow angle. This induced power is decreasing as the rotor becomes more efficient (translational lift).

Total power is continuing to decrease with increasing airspeed until you reach the "bucket" speed. This is the point of greatest difference between power required and power available, translating into a maximum climb rate. Beyond this speed, the rotor continues to become more efficient. Still, wind resistance begins to prevail, and parasite power swaps with induced power as the main contributor to total rotor power required. Total power then begins to rise until it meets with power available, defining the helicopter's maximum horizontal speed.

Although the blades and airframe move easier through the thinner air of higher altitudes, more pitch in the blades required creating thrust, and an overall increase in total power required can be expected with altitude increase.

Having a keen understanding of your aircraft's power requirements can help you perfect technique, maximize performance, and minimize maintenance. Induced power will decrease with a greater airflow through the rotor disk in a straight forward flight. Similarly, in a max performance vertical takeoff — a pilot can finesse a heavy aircraft into the air by watching their power margin increase on a decreasing torque meter allowing more "pitch pull," while still staying within limits. Knowing that left pedal or a left cyclic roll will drive power requirements up on rotors rotating clockwise (looking up), a pilot can avoid an over-torque by not operating too close to a limit while maneuvering. The opposite pedal/cyclic will, of course, do the same on opposite spinning rotors.

# Power failure at hoover

Multi-engine helicopters do bring even more the need to understand power, especially in a one-engine-inoperative flight that requires their own discussion making. It's sufficed to say that power may be important for helicopters, but knowledge will always reign.

Always operate the helicopter as close to its normal capabilities as possible, considering the situation in hand. Except for the pinnacle operation, the angle of descent should be no steeper than necessary to clear a barrier in the approach path and still land on the selected spot in all confined area operations. Clearing a barrier by a few feet and maintaining normal operating **RPM**, with perhaps a reserve of power, better than clearing a barrier by a wide margin but with a dangerously low **RPM** and no power reserve.

Always make the landing to a specific point and not to some general area. This point is located further from the approached area. The more confined the area, the more essential it is to land the helicopter precisely at a definite point. Keep this point of insight during the entire approach.

When flying a helicopter near obstructions, always consider the tail rotor. You must establish A safe angle of descent to ensure the tail rotor clearance of all obstructions. After coming to a hover, take care to avoid turning the tail into obstructions.

# Low rotor speed

Low Rotor RPM in a helicopter is conditioned, the automatic response of rolling on the throttle and lowering collective. Catastrophic blade stall is fatal and a result of allowing the main rotor RPM to decay to no recovery point. You would notice a decrease in engine speed and a rotor noise, a slight vibration, and cyclic stick shake at higher speeds. Next is the activation of the low rotor RPM warning light and the buzzer system. There's a safety factor built into helicopters, any time the rotor speed falling below the green arc and the power, simultaneously adding throttle and lowering the collective. If in forward flight, you gently apply aft cyclic to cause more airflow through the rotor system and increase the rotor speed. If you are left without power, immediately lower the collective and apply aft cyclic. This is a low rotor speed situation.

Helicopter pilots have a saying: speed= life. It means that if your blades are spinning fast enough, you should be able to fly. But if you lose engine speed, there's a chance that you might drop out of the sky like a brick and have a very ugly encounter with the ground. Why? Because the spinning of the rotor blades is what gives a helicopter lift. If they stop spinning, they're not generating lift. If they're not spinning fast enough, they're not generating enough lift to keep the helicopter airborne.

Helicopters have low rotor RPM warning systems. In an R44, it consists of light on the instrument panel and a horn or a buzzer.

The sound of this horn very annoying and impossible to miss. Because RPM is so important, the full system lights and horn are required for flight.

On a Robinson helicopter, the low rotor RPM warning system kicks in at 97% RPM. Since the helicopter operating at 102% RPM, that's just 5 units below normal operation. But as they teach in the Robinson Safety Course, the helicopter should fly with RPM of 80% + 1% per 1,000 feet of density altitude. It is not recommended to fly a helicopter below normal operating RPM. This rule of thumb will help you understand how critical a low rotor RPM situation might be.

# Hazardous Conditions

Pilots should be aware of some dangerous circumstances that they may encounter. The following is requiring an immediate and correct response to avoid losing control. Engine failure, tail rotor failure, and stalling of the rotor blades due to low rotor speed also require immediate corrective measures.

Turbulent air that gets near the tail rotor can cause yaw instability as well. The constantly changing airflow will create a fluctuating lift from the rotor requiring power change to maintain a hover.

A heavy helicopter that operates at a high-density altitude, insufficient available power could cause them to land hard.

However, a bigger danger actually striking an object with the rotor system. According to the NTSB, a Sikorsky S-76 taking off from a hospital helipad became distracted by a large object, flapping in the wind. The main rotors struck the top corner of the building, and the helicopter began to settle.

The pilot lowered the collective and navigated between the building and the parking garage to the street below, applying full collective pitch to help cushion the hard landing. The pilot and two medical crewmembers received minor injuries.

Another pilot was not alerted to anything unusual until he looked up and noticed the helicopter's close proximity to a 16-floor brick building that extended above the helipad's height by four floors. The pilot made a rapid right cyclic input to avoid hitting the building.

But the helicopter struck the building and fell about 13 floors to ground level.

We learn that operating in a confined area during gusty wind conditions requires enhanced concentration, especially during operations at high-density altitudes.

# Gyroscopic Precession

The gyroscopic Precession principle states that when a force is applied to a spinning object, the maximum reaction occurs approximately 90 degrees later in the rotation direction. Since a helicopter's main rotor acts as a gyroscope and has gyroscope action properties, one of them is precession. Gyroscope precession is the deflection of any spinning object when a force applied to it. This occurs at about 90° in the direction of gyration from where the force applied. The blades tend to flap up or down as the angle of incidence increased or decreased.

You need to know how a rotor disc tilts in the direction commanded by the pilot. Each blade's pitch is controlled by a swashplate, and each blade is connected to the upper ring of the swash plate through a pitch link. The cyclic flight control is connected to the lower ring. When inputs are made, it tilting the swashplate and changing the pitch of each blade independently. This can be observed during a pre-flight. If you move the cyclic While the swash plate actually tilts in the direction that the cyclic control moves, each blade's pitch must change approximately 90 degrees before getting the rotor disk to also tilt in that direction. This is accomplished by the blades' pitch horns, which are offset approximately 90 degrees. Since maximum deflection takes place approximately 90 degrees later, the disc tilts forward.

# Forced landing

The approach is the same as a normal approach to a hover. However, instead of terminating at a hover, continue the touchdown approach.

Touchdown should occur with the skids level, zero ground speed, and a rate of descent approaching zero.

As the helicopter comes near the surface, you need to increase the Collective, as necessary, to cushion the landing on the surface and terminate a skids-level altitude without straight forward movement. You should also consider a forced landing area in case of an emergency.

A high reconnaissance should fly at an altitude of 300 to 500 feet above the surface. As a general rule, ensure that sufficient altitude is available at all times to land into the wind in case of engine failure. Always maintain safe altitudes and airspeeds, and keep a forced landing area within reach whenever possible.

# Magnetic course

Magnetic course (MC) the True Course (TC) adjusted for magnetic variation. The magnetic course the course that should take you directly from origin to destination if there were no wind along the route of flight.

Determine the True Course (TC) by placing the straight edge of a navigational plotter or the protractor along the route, with the plotter's hole on the intersection of the route and a meridian of longitude (the vertical line with small tick-marks).

The TC is measured by the numbers on the protractor portion of the plotter at the meridian line. Note that up to four numbers (90° apart) are provided on the plotter. You need to determine the direction of the flight, using common sense.

Alternatively, you could use a latitude (horizontal line with small tick-marks) if your course in a north or south direction. This why there four digits in the plotter.

You may use either a meridian or line of latitude to measure your course and going in either direction along the course line.

To determine the Magnetic Course (MC) for your flight route, adjusted the True Course (TC) by adding or subtracting the charted magnetic variation. On sectional charts, a long-dashed line provides the number of degrees of magnetic variation (angle between true north and magnetic north).

The variation in either East or West and signified by "E" or "W," for example, 3°E or 5°W. If the variation east, subtract the variation from the TC; if the variation west, add the variation to the TC. Use this saying as a memory aid: "East least. West best."

If you determine your course using the compass rose around a VOR, the compass rose already adjusted for the magnetic variation, so your course already magnetic, and no adjustment is needed from TC to MC. The compass rose surrounds VORs and has every 30° labeled and tick marks for every 5°.

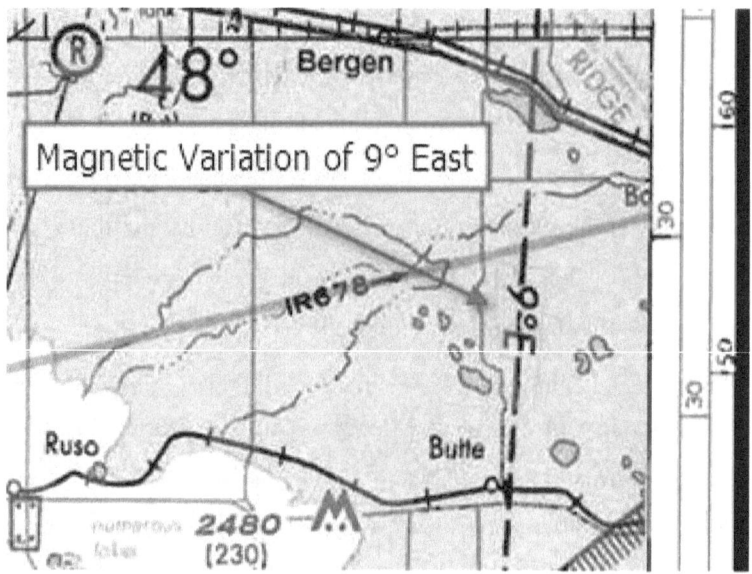

Remember to use the reciprocal heading of radials when flying TO the VOR. For example, if your flying course is toward an airport on the 090° radial, your heading is towards the airport, which is MC 270°, and not 090°.

Computer-generated forecast charts of winds and temperatures aloft are available for international flights at specified levels. The U.S. National Centers for Environmental Prediction (NCEP), near Washington, D.C., prepares and supplies to users' charts of forecast winds, temperatures, and significant weather.

# Compass Heading

Compass Heading Magnetic Heading (MH), corrected for compass deviation. The compass's deviation is because it is installed in an aircraft - all metal parts in the airplane affect the compass, causing deviation. The deviation of each compass in an aircraft is given on the Compass Correction Card. The card is usually mounted on the compass itself, with correction intervals for every 30°. For each 30°, the card lists both the MH and its corresponding CH (sometimes simply listed as a "steer" heading; since these the headings, you should "steer" to arrive at the destination.

If a question asks for a compass heading, the Magnetic Heading is converted to a CH by adding or subtracting deviation. The difference between the two headings listed on the Compass Correction Card the deviation, the amount to add, or subtract to Magnetic Heading to arrive at CH. Though they are not required for flight, it very easy and prudent to file a VFR flight plan for any cross-country type of flight (i.e., when you plan to fly from one airport to another).

• Cruising Altitude only your initial requested altitude.

• The airport's destination or place where you plan to make your last landing of this flight.

• Unless you plan a stopover of more than 1 hour elsewhere, En-route.

• Fuel on Board

Fuel on Board the amount of usable fuel in the airplane (listed in hours and minutes of flying time) at the time of departure.

A VFR flight plan does not close or cancel itself - it has to close actively. Control towers do not close VFR flight plans as you land. Close your flight plan with the nearest FSS. If FSS not available, you may request any ATC facility to relay your cancellation to the FSS.

Aircraft manufacturers are responsible for generating flight performance data, which flight planners use to estimate fuel needs for a particular flight—the fuel burn rate based on specific throttle settings for climbing and cruising. The planner uses the projected weather and aircraft weight as inputs to the flight performance data to estimate the necessary fuel to reach the destination. The fuel burn gave fuel (usually pounds or kilograms) instead of the volume (such as gallons or liters) because of aircraft weight critical. In addition to standard fuel needs, some organizations require that a flight plan include reserve fuel if certain conditions are met.

For example, an over-water flight of longer than a specific duration may require the flight plan to include reserve fuel.

The reserve fuel may plan as an extra, which leftover on the aircraft at the destination, or it may assume to burn during the flight (perhaps due to unaccounted for differences between the actual aircraft and the flight performance data).

In case of an in-flight emergency, it may be necessary to determine whether it quicker to divert to the alternate airfield or continue to the destination.

Contact Information at Destination: Having a means of contacting you is useful for tracking down an aircraft that has failed to close its flight plan and possibly overdue or in distress.

# Area Forecasts

Forecasts of general weather conditions over several states do not contain forecasts of the winds and temperatures aloft. The International Weather Depiction Chart indicates current weather and does not forecast winds and temperatures aloft.

The Low-Level Prognostic Chart depicts weather conditions that forecast to exist at a specific time shown on the chart. Prognostic charts forecast conditions, not report observed conditions (as the Weather Depilation Chart does). Low-level Prognostic Charts forecast conditions 12 and 24 hours (not 6 hours) after the time of issuance.

# FLIGHT OPERATION

# Pressure altitude

Helicopters are still susceptible to altitude-related issues since, even at an elevation of 1500 feet, atmospheric pressure decreases to 13.9 psi. The rotors push air downwards, allowing the chopper to move upwards against the force of gravity. As it turns out, the lift generated by the helicopter rotors depends on several factors, and the density of air is one of them. High-performance helicopters like the AgustaWestland AW109 can hover at 10,400 feet.

Turbine-engine helicopters can reach around 25,000 feet. But the maximum height at which a helicopter can hover is much lower - a Pressure altitude indicates when the altimeter is set to the standard sea level pressure (29.92" Hg). In the United States, altimeters are always set to 29.92" Hg at and above 18,000 feet.

Both true and pressure altitudes should the same at FL310 if the ambient air temperature was standard. Pressure altitude should be lower than the true altitude and warmer than the standard air temperature. The pressure is measured at a station or airport 'station pressure' or the actual pressure at field elevation. The altimeter setting the value to which the scale of a pressure altimeter adjusted to read field elevation. Station barometric pressure reduced to sea level is a method to compare station pressures between stations at different elevations.

If the Pitot tube becomes blocked, but the pressure is not trapped in the pitot lines, the indicated airspeed will drop to zero if the ram air input is blocked since the pitot pressure will approximately equal to the static pressure. The airspeed line pressure will vent out through the hole, and the indication will drop to zero.

# Flight Routes

The aviation authorities in the country are in control of flight routes. Beacons signal-emitting devices before radar and other sophisticated equipment on planes now a day. The flight routes of civilian planes stick to those routes.

The GPS systems on the plane assist with the autopilot system.

Most flights will always follow the same route unless the route is unavailable. The flight plan may change to another flight route due to air traffic or weather conditions.

The PILOT cannot change flight En-route without the authority of the specific ATC who controls the routes.

Most flight routes will avoid the minimum time over the sea and keep close to an airport in case of an emergency.

# Flight levels

Flight levels (FL), used by Air Traffic Control to simplify the vertical separation of aircraft, exist every 1000 feet relative to an agreed pressure level.

Above a transitional altitude, the worldwide arbitrary pressure datum of 1013.25 millibars or the equivalent setting of 29.92 inches of mercury entered into the altimeter, and altitude then referred to as a flight level.

The altimeter reading converted to a flight level by removing the trailing two zeroes: 29000 feet FL290.

When the pressure at sea level by chance the international standard, then the flight level also the altitude. Airways have associated standardized flight levels known as the flight model, which must be used when on the airway. On a bi-directional airway, each direction has its own set of flight levels. A valid flight plan must comprise a legal flight level at which the aircraft will travel the airway. A change in the airway may require a change of flight level.

In the USA and Canada, for eastbound a heading 0–179 degrees IFR flights, the flight plan must list as an 'odd' flight level in 2000-foot increments starting at FL190 (i.e., FL190, FL210, FL230, etc.).

Westbound (heading 180–359 degrees) IFR flights must list an 'even' flight level in 2000-foot increments starting at FL180 (FL180, FL200, FL220, etc.).

The Air Traffic Control (ATC) may assign any flight level at any time if traffic situations merit a change in altitude.

The maximum height at which a helicopter can hover is much lower - a high-performance helicopter like the Agusta A109E can hover at 10,400 feet. The Agusta can hover in ground effect - at 13,800 feet.

As helicopters become faster and more agile, pilots are expected to navigate at low altitudes while traveling at high speeds. A pilot's ability to interpret information from a combination of visual sources determines mission success and the aircraft and crew survival.

The highest altitude of the Sikorsky R-4 is about 12,000 feet.

The highest altitude of the Bell 47 is about 18,550 feet.

The highest altitude of Aerospatiale SA-313 Alouette is 26,932 feet.

The highest altitude of the Bell UH-1 Iroquois is about 14,500 feet.

The highest altitude records of the Mil Mi-8 are about 30,000 feet.

The highest Altitude of the Boeing CH-47 Chinook is 18000 feet.

The highest Altitude of the Sikorsky H-60 Black Hawk is about 19,151 feet.

The highest Altitude records of the AS350 are about 29,000 feet.

# Headwinds effect

An increase in headwind component (which could also cause by a tailwind shearing to calm, causes airspeed and pitch to increase, sink rate to decrease. A sudden decrease in the headwind component should decrease aircraft performance, and it should indicate by the decrease in airspeed, pitch, and altitude.

An increase in tailwind velocity should decrease performance and indicate a decrease in airspeed, pitch, and altitude.

When a headwind shears to calm or a tailwind, the aircraft tends to lose airspeed, get low, and pitch nose down. The aircraft will require more power and a higher-pitch altitude to stay on the glide-slope. With a headwind shearing to a calm wind, there a loss of lift as airspeed decreases, the aircraft pitches down, and the aircraft drops below glideslope (altitude decreases).

Responding promptly by adding power and pitching up, a Helicopter pilot may overshoot the glide slope and airspeed target but then recover. The aircraft will pitch down due to the relatively small angle of attack used during the headwind and the sudden decrease in the airflow over the wing when the wind shears calm.

A headwind increasing against the pitot and airframe will result in an airspeed increase.

Less power required maintaining an indicated airspeed in a headwind than in calm air because of ram air; thus, a shear from a headwind to calm should be indicated by decreased airspeed and a decrease in altitude.

A headwind increasing against the pitot and airframe will result in an airspeed increase.

# Traffic Patterns

A traffic pattern is useful to control the flow of traffic, mainly at uncontrolled airports. Airplanes and helicopters do not mix well in the same traffic situation. Helicopters should always avoid the flow of fixed-wing traffic. Familiar with the patterns flown by airplanes. Learn these patterns in case ATC requests a fixed-wing traffic pattern flown.

A standard helicopter traffic pattern is rectangular and comprises of right turns, flown at 500-1,000 feet AGL depending on consideration.

A standard pattern when all turns are made to the left. Begin the base leg at a selected point and begin the turn sooner than normal if the wind is strong. Delay the turn-to-base if the wind is light. Flying a traffic pattern should not a problem for a helicopter unless its maneuvers require specific altitudes.

When a helicopter approaching an airport with ATC, adhere to the standard practices and patterns. Generally, helicopters make a lower altitude pattern opposite from the fixed-wing pattern and make their approaches to some point other than the runway.

The standard departure procedure is typically straight-out, downwind, or right-hand departure when using the fixed-wing traffic pattern.

If a control tower is in operation, request the type of departure desired.

Mostly, helicopter departures are made into the wind unless it's dictated otherwise. In uncontrolled airports, comply with the departure procedures for that airport.

As the helicopter nears its final approach path, the final decision should be made depending on the obstructions and the forced landing areas. The landing area should always be in your sight, and the angle of approach should never be too high or too low to the landing surface.

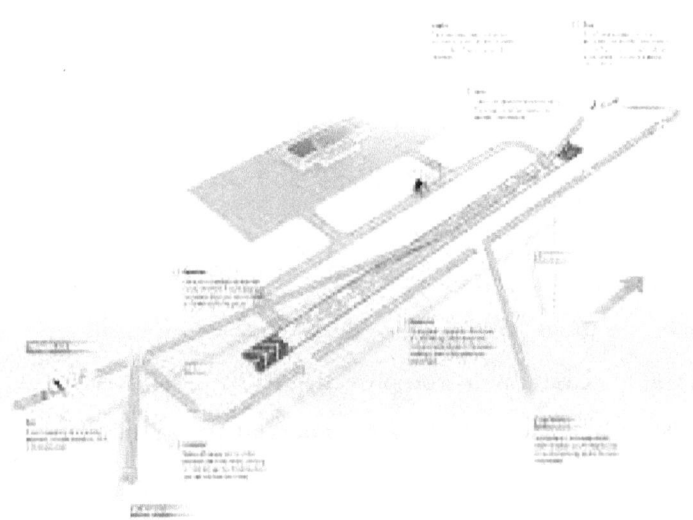

# Descent

To descend at a comfortable rate without building too much speed, you must decrease the main rotor pitch by lowering the Collective. Anticipate the need for the right force pedal as you decrease power.

The nose will drop as you lower the Collective so that you need to add a little aft Cyclic so to maintain the correct pitch attitude and the airspeed. You should not add too much aft Cyclic, or the aircraft will climb.

Note that as you descend, the engine will produce more power. Monitor the engine instruments and smoothly reduce the Collective to continue the descent.

To level off from a descent, you should start increasing the Collective about 50 feet above the altitude you want to level off. As you increase power, you press the left force pedal and use the Cyclic to maintain cruising airspeed. You then apply a forward Cyclic to increase the speed and aft Cyclic to slow down.

# Visual flight rules(VFR) Flights

Operating under Visual flight rules(VFR) Flight, you fly solely by reference to the outside visual cues, such as the horizon, buildings, etc., which permit navigation and separation from terrain and other traffic. The cloud ceiling and flight visibility the most important for safe operations during flight. The minimum weather conditions for ceiling and visibility for **VFR** flights are defined in **FAR** Part 91.155, depending on the airspace type. A typical daytime **VFR** minimum for most airspace 3 statute miles of flight visibility and a cloud distance of 500' below, 1,000' above, and 2,000' feet horizontally. Visual flight rules are simpler than the **IFR** flight and require significantly less training and practice.

VFR Helicopter pilots may use cockpit instruments as secondary the aids to the navigation and orientation, but it's not required to do so.

Any aircraft operating under VFR must have the required equipment on board, as described in FAR Part 91.205, which comprises instruments necessary for IFR flight, but the view outside of the aircraft the primary source for keeping the aircraft straight and level, flying without hitting anything.

# Instrument flight rules (IFR)

Instrument flight rules (IFR) permit any aircraft to operate in instrument meteorological conditions (IMC) in contrast to VFR. They also an integral part of flying in class A airspace. 'Class A' airspace exists over and near the 48 contiguous U.S. states and Alaska from 18,000 feet above mean sea level to FL 600. Flight in 'class A' airspace has required the pilots to be rated to operate under Instrument Flight Rules (IFR).

The aircraft must be equipped with an IFR instrument. Most of the jet aircraft operating in 'Class A' Airspace for the cruise portion of their flight required to utilize IFR procedures.

# NIGHT FLIGHT

Human eyes' functions are not effective at night as animals' eyes with nocturnal habits. Therefore, it is important to understand the eye's function and the effect of darkness. Innumerable light-sensitive nerves, called "cones" and "rods," are located at the back of the eye or retina, where all images are focused. These nerves connect with the cells of the optic nerve, which transmits messages directly to the brain. The cones are located in the center of the retina. The rods are concentrated in a ring around the cones of the eyes.

Both the cones and the rods are used for vision during daylight. The rods and cones function in daylight and in the moonlight.

But in the absence of normal light, the process of night vision is placed almost entirely on the rods.

You should consciously practice this scanning procedure to improve night vision. The eye's adaptation to darkness is another important aspect of night vision. When entering a dark room, it is difficult to see until the eyes are adjusted to the darkness. After approximately 5 to 10 minutes, the cones adjusted to the dim light, and the eyes become 100 Cones for much more time. About 30 minutes is needed for the rods to become adjusted to the darkness, but they are about 100,000 times more sensitive to light than they were in the lighted area when they do adjust.

After the adaptation process is completed, much more could be seen, especially if the eyes are used correctly. The entire process is reversed when entering a lighted room. The eyes are first dazzled by the brightness, but become completely adjusted in a very few seconds, thereby losing their adaptation to the dark.

If the darkroom is reentered, the eyes go through the process of adapting to the darkness again.

After the eyes have become adapted to the darkness, the Pilot should avoid exposing them to bright white light.

That will cause temporary blindness and could result in serious consequences.

Temporary blindness, caused by an unusually bright light, may result in illusions or after images until the eyes recover from the brightness. This results in misjudging or incorrectly identifying objects, such as mistaking slanted clouds on the horizon or populated areas of a landing field.

Vertigo is experienced as a feeling of dizziness and imbalance that could create or increase illusions. The illusions seem very real, and Pilots at every level of experience and skill could be affected. Recognizing that the brain and eyes could play tricks is the best protection for flying at night.

In addition to night vision limitations, Pilots should be aware that night illusions could cause confusion and concerns during night flying. On a clear night, distant stationary lights could be mistaken for stars or other aircraft.

Dark nights tend to eliminate the reference to a visual horizon, and as a result, you need to rely less on outside references at night and more on the flight and navigation instruments. Visual auto kinesis could occur when you stare at a single light source for several seconds on a dark night.

This result of light appears to be moving. The auto kinesis effect will not occur if you expand the visual field.

It is a good procedure not to become fixed on one source of light. You should try to eliminate any light source causing blinking or flickering problems in the cockpit.

A black-hole approach occurs when the landing is made from over water or non-lighted terrain, where the runway lights are the only source of light. Without peripheral visual cues to help, most pilots will have trouble orientating themselves relative to Earth. The runway could seem out of position (down sloping or upsloping) and, in the worst case, result in landing short.

If an electronic glide slope or visual approach slope indicator (VASI) is available, it should be used. If navigation aids (NAVAIDs) are unavailable, careful attention should be given to using the flight instruments to assist in maintaining orientation and a normal approach.

Bright runway and approach lighting systems, especially where few lights illuminate the surrounding terrain, may create the illusion of less distance to the runway. In this situation, the tendency is to fly a higher approach.

With this situation, the tendency is common to fly at the lower-than-normal approach. If the runway has a city in the distance on higher terrain, the tendency will be flying at the lower-than-normal approach.

When a double row of approach lights joins the runway's boundary lights, there could be confusion where the approach lights terminate, and runway lights begin.

Aircraft position lights are arranged similarly to those of boats and ships. A red light is positioned on the left wingtip, a green light on the right wingtip, and a white light on the tail.

This arrangement provides a means by which you could determine the general direction of another aircraft movement in flight.

If both a red and green light of other aircraft were observed, the aircraft should fly toward the Pilot and be on a collision course.

Landing lights are not only useful for taxi, takeoffs, and landings but also provide a means by which aircraft could be seen at night by other Pilots.

You are encouraged to turn on the landing lights when operating within 10 miles of an airport for day and night or in reduced visibility.

This should also be done in areas where flocks of birds may be expected. Most aircraft lights blend in with the stars or the cities' lights at night and go unnoticed unless a conscious effort is made to distinguish them from other lights. The lighting systems used for airports, runways, obstructions, and other visual aids at night are other important aspects of night flying.

Lighted airports located away from congested areas could be identified readily at night by the lights outlining the runways.

Airports located near or within large cities are often difficult to identify in the maze of lights.

It is important not only to know the exact location of an airport relative to the city but also to identify these airports by their lighting pattern characteristics. Aeronautical lights are designed in a variety of colors and configurations, each having its own purpose. Before a night flight, particularly a cross-country night flight, the Pilot checks the lighting systems' availability and status at the destination airport.

You can find this information on aeronautical charts and in the Airport/Facility Directory.

The beacon rotates at a constant speed, thus producing what appears to be a series of light flashes at regular intervals. These flashes may be one or two different colors used to identify various landing areas.

Beacons producing red flashes indicate obstructions or areas considered hazardous to aerial navigation.

Steady burning red lights are used to mark obstructions on or near airports and sometimes supplement flashing lights on En-route obstructions.

High intensity flashing white lights mark some supporting structures of overhead transmission lines that stretch across rivers, chasms, and gorges.

These high-intensity lights are also used to identify tall structures, such as chimneys and towers. Because of the technological advancements in aviation, runway lighting systems have become quite sophisticated to accommodate takeoffs and landings in various weather conditions. The Pilot whose flying is limited to VFR only needs to be concerned with the following basic lighting of runways and taxiways.

Prominently lighted checkpoints along the prepared course should be noted. Rotating beacons at airports, lighted obstructions, lights of cities or towns, and lights from major highway traffic all provide excellent visual checkpoints.

# Visual illusions

Visual illusions are a familiar factor for most of us. Even under conditions of good visibility, one can experience visual illusions. This illusion may make a pilot change (increase or decrease) the slope of his final approach. They are caused by runways with different widths, up sloping or down-sloping runways, and up sloping or down sloping in final approach terrain. Pilots learn to recognize a normal final approach by developing and recalling a mental image of the expected relationship between the length and the width of an average runway. An example would be a pilot used to small general aviation fields visiting a large international airport.

The much wider runway would give the pilot the mental picture of the point where he/she would usually begin the flare when they are much higher than they should be. A pilot flying an aircraft where the cockpit height relative to the ground is vastly higher or lower than they are used to can cause a similar illusion in the last part of the approach.

A final approach over an up-sloping terrain with a flat runway, or to an unusually narrow or long runway may produce the visual illusion of being too high on final approach. The pilot may then increase his descent rate, positioning the aircraft unusually low on the approach path.

A final approach over a down-sloping terrain with a flat runway, or to an unusually wide runway may produce the visual illusion of being too low on final approach. The pilot may then pitch the aircraft's nose up to increase the altitude, resulting in a low-altitude stall or a missed approach.

# Black-hole approach illusion

A black-hole approach illusion can happen during the final approach at night (with no stars or moonlight) over water or unlit terrain to a lighted runway, in which the horizon is not visible. Pilots confidently proceed with a visual approach instead of relying on instruments during nighttime landings. As a result, this can lead to the pilot experiencing glide path overestimation. As a result, they initiate an aggressive descent and wrongly adjust to an unsafe glide path below the desired three-degree glide path.

# Autokinetic illusion

The Autokinetic illusion occurs at night or in conditions with poor visual cues. This illusion gives the pilot the impression that a stationary object moves in front of the airplane's path; it is caused by staring at a single fixed point of light (ground light or a star) in a dark and featureless background. The reason this visual illusion occurs is very small movements of the eyes. In conditions with poor visual cues accompanied by a single source of light, the brain's eye movements are interpreted as the object's movement. [6] This illusion can cause a misperception that such a light is on a collision course with the aircraft.

Planet or stars in the night sky can often cause the illusion to occur. These bright stars or planets have often been mistaken for landing lights of oncoming aircraft, satellites, or even UFOs. An example of a star that commonly causes this illusion is Sirius, the brightest star in the northern hemisphere. It appears over the entire continental United States at one to three fists widths above the winter horizon. At dusk, the planet Venus can cause this illusion to occur, and many pilots have mistaken it as lights coming from other aircraft

# AIR TRAFFIC
# CONTROL (ATC)

Air traffic control (ATC) a service provided by ground-based controllers who direct aircraft on the ground and through controlled airspace and can provide advisory services to aircraft in non-controlled airspace. ATC's primary purpose is to prevent collisions, organize and expedite traffic flow, and provide information and other support for Helicopter pilots. To prevent collisions, ATC enforces traffic separation rules, which ensure that each aircraft maintain a minimum amount of empty space around it at all times.

ATC may issue instructions that Helicopter pilots are required to obey or advisories that Helicopter pilots may, at their discretion, disregard it.

You are the final authority for the safe operation of the aircraft. In an emergency, you may deviate from ATC instructions to the extent required to maintain their aircraft's safe operation.

According to the International Civil Aviation Organization (ICAO) requirements, ATC operations are conducted either in the English language or in the station's language on the ground. In practice, the native language of a region is normally used.

However, the English language must use upon request—the primary method of controlling the immediate airport environment visual observation from the Aerodrome control tower.

Tower controllers responsible for the separation and efficient movement of aircraft and ground vehicles, operating on the taxiways and runways of the airport itself, and aircraft in the air around the airport, generally 5 to 10 nautical miles, depending on the airport procedures.

Surveillance also displays available to controllers at larger airports to assist with controlling air traffic. Controllers may use a radar system known as Secondary Surveillance Radar for airborne traffic approaching and departing.

These displays comprise a map of the area, the position of different aircraft, and data tags that comprise aircraft identification, speed, altitude, and other information described in local procedures.

In adverse weather conditions, the tower controllers may also use Surface Movement Radar, Surface Movement Guidance, and Control Systems (SMGCS).

Ground Control is responsible for the airport's movement's areas and areas not released to the airlines or other users. This comprises all taxiways, inactive runways, holding areas, and some transitional intersections where the aircraft arrive, having vacated the runway or departure gate.

Any aircraft, vehicle, or person walking or working in these areas must have clearance from Ground Control.

This is normally done via VHF/UHF radio. Aircraft or vehicles without radios must respond to ATC instructions via aviation light signals or else led by vehicles with radios.

People working on the airport surface normally have a communications link to communicate with Ground Control. Ground Control vital to the smooth operation of the airport. Some airports have Surface Movement Radar (SMR), such as ASDE-3, AMASS, or ASDE-X, designed to display aircraft and vehicles on the ground.

These are used by Ground Control as an additional tool to control ground traffic.

Tower Control responsible for the active runway surfaces. Tower Control clears aircraft for takeoff or landing, ensuring that prescribed runway separation will exist at all times. If Local Control detects any unsafe condition, a landing aircraft may tell to 'go-around' and re-enter into the approach or terminal area controller's landing pattern.

When there an extremely high demand for a certain airport or the airspace nearby becomes congested, there may ground 'stops' or re-routing necessary to ensure the system does not get overloaded.

Flight Data may inform the Helicopter pilots using a recorded continuous loop on a specific frequency known as the Automatic Terminal Information Service (ATIS).

Traffic flow is broadly divided into departures, arrivals, and overflights. As an aircraft move in and out of the terminal airspace, they handed off to the next appropriate control facility.

ATC controller is responsible for ensuring that aircraft at an appropriate altitude when they are handed off and that aircraft arrive at a suitable lending rate.

Not all airports have a radar approach or terminal control available. At some of these airports, the tower may provide a non-radar procedural approach to arriving aircraft that handed over from a radar unit before they could to land.

ATC provides services to aircraft in flight between airports as well. Helicopter pilots fly under two sets of separation rules: Visual Flight Rules (VFR) or Instrument Flight Rules (IFR). Air traffic controllers have different responsibilities to aircraft operating under the different sets of rules. En-route air traffic controllers issue clearances and instructions for airborne aircraft, and Helicopter pilots must comply with these instructions. Controllers in a center adhere to separation standards that define the minimum distance allowed between aircraft.

Each center responsible for many thousands of square miles of airspace and for the airports within that airspace. Centers control IFR aircraft from when they depart from an airport or terminal area's airspace to the time they arrive at another airport or terminal area's airspace.

Center controllers responsible for climbing the aircraft to their requested altitude while, at the same time, ensuring that the aircraft properly separated from all other aircraft in the immediate area.

As an aircraft reaches the boundary of a Center's control area, it handed over to the next Control Center. After the handover, the aircraft gave a frequency change, and you begin talking to the next controller. This process continues until the aircraft handed off to ATC for the approach.

Centers will typically use long-range radar that can see aircraft within 200 nautical miles of the radar antenna. They may also use the TRACON radar data when it provides better piloting of the traffic.

The centers also exercise control over traffic traveling over the world's ocean areas. Because there no radar systems available for oceanic control, oceanic controllers provide ATC services using procedural control. These procedures use aircraft position reports, time, altitude, distance, and speed to ensure separation. This process requires that greater distances for any given route separate the aircraft.

Precision approach radars are commonly used by military controllers of the Air Forces of several countries. He was done to assist you in the final phases of landing, in places where the Instrument Landing System (ILS) or other sophisticated airborne equipment was unavailable to assist him in near-zero visibility conditions.

A Radar Archive System (RAS) keeps an electronic record of all radar information, preserving it for a few weeks. This information can useful for search and rescue.

When an aircraft has 'disappeared' from radar screens, a controller can review its latest radar returns to determine its likely position.

All ATC facilities using radar weather processors with an ability to determine precipitation intensity to describe the intense to Helicopter pilots such as LIGHT (< 30 DBZ), MODERATE (30 to 40 dB), HEAVY (>40 to 50 dB), or EXTREME (>50 DBZ). The designation 'SPECI' means that this a special weather observation.

# ATC operation

ATC will issue landing clearances to helicopters going to movement areas other than active runways or from diverse directions to points on active runways, with additional instructions as necessary. Unless agreed to by the helicopter pilot, ATC will not issue takeoffs downwind if the tailwind exceeds 5 knots. As a helicopter pilot, you should request to takeoff from a given point in a given direction that constitutes an agreement.

Separating a departing helicopter from other helicopters is done by ensuring that it does not take off until one of the following conditions exists:

- Helicopters performing air-taxiing operations within the boundary of the airport considered taxiing aircraft.

- A preceding, departing helicopter has left the takeoff area.

- A preceding, arriving helicopter has taxied off the landing area.

Authorize helicopters to conduct simultaneous landings or takeoffs if the distance between the landing and takeoff points at least 200 feet and the courses to flown do not conflict. The ATC will instruct a helicopter to remain at least 200 feet from another helicopter.

# RADIO COMMUNICATIONS

# Radio communication

If a two-way radio communication failure occurs while in IFR conditions, you should continue the flight by the following route:

• By the route assigned in the last ATC clearance;

• If being radar vectored, by the direct route from the point of the radio failure to the fix, route or airway specified in the vector clearance;

• In the absence of an assigned route, flying the route that ATC has advised may expect in a further clearance; or

• In the absence of an assigned route or a route that ATC has advised may expect in a further clearance by the route filed in the flight plan.

The route shown on the flight plan should be the last route to be used and only if an assigned route, vector, or expected route has not been received. A climb should only be initiated to establish the highest of the assigned MEA or the expected altitude. The squawk of 7700 is no longer correct. TIBS provides continuous telephone recordings of meteorological and aeronautical information. It also provides specific area and route briefing and airspace procedures and special announcements, if applicable.

It is designed as a preliminary briefing tool and cannot replace a standard briefing from a flight service specialist.

TIBS is available 24 hours a day by calling 1-800-WS-BRIEF, and it's updated when conditions change. As a minimum, area briefings encompass a 50 NM radius.

If no EFC time has been received, commence descent and approach are as close as possible to the estimated arrival time as calculated from the filed or amended (with ATC) estimated time En-route.

Radio communications a critical link between Helicopter pilots and Air Traffic Control. The pilots must acknowledge each radio communication with ATC by using the appropriate aircraft call sign. The communication should be as brief as possible. Still, Air Traffic Control must know what you want to do before monitoring radio communications frequencies with the aircraft.

Good phraseology enhances safety and is the mark of a professional helicopter pilot. Many times, you can get the information through ATIS by monitoring their frequency.

If you hear someone else talking, keying the transmitter will be futile, and you will probably jam their receivers, causing them to repeat their call.

When you change frequencies, you should pause, listen, make sure the frequency is clear, and then speak in a normal, conversational tone.

If there a lack of sounds in the receiver, you should check the volume, frequency and make sure that the microphone is not stuck in the transmit position. This type of interference is referred to as a 'Stuck Mike.'

Remote radio sites do not always transmit and receive on all of the available frequencies, particularly concerning VOR sites where you can hear but not reach a ground station's receiver.

The initial call-up is the first radio call you make to a given facility or the FSS specialist within a facility.

Many FSSs are equipped with Remote Communications Outlets RCOs and can transmit on the same frequency at more than one location. The frequencies are available on charts above FSS communications boxes. If the chart indicates, the FSS frequencies are above the VORTAC or in the FSS communications boxes, transmit or receive frequencies nearest to the Helicopter pilot location.

Sometime the Air Traffic Controller must issue a time-critical instruction to other aircraft and may not be in a position to respond to your call immediately.

When you advised by ATC to change frequencies, you must acknowledge the instruction.

Suppose you have selected a new frequency without acknowledgment. In that case, the Air Traffic Control's workload will increase because there no way of knowing whether you received the instruction or have had radio communications' failure.

You must select the new frequency as soon as possible unless instructed to change at a specific time or altitude. A delay in making the changes could result in untimely receipt of important information.

Civil aircraft Helicopter pilots should state the aircraft type, model, or manufacturer's name, followed by the registration number's digits/letters.

Helicopter pilots must be certain that the aircraft identification is complete and identified before taking action on an ATC clearance.

Air carriers, who have FAA authorized call signs, should identify themselves by stating the complete call sign using group form for the numbers, and the word 'HEAVY' if appropriate.

When an air carrier dispatches a flight using another company's equipment, and you do not advise the terminal ATC facility, the possible confusion in aircraft identification can compromise safety.

ATC facilities may also request helicopter pilots to use letter equivalents when aircraft with similar-sounding identifications on the same frequency. Helicopter pilots should use the phonetic alphabet when identifying their aircraft during initial contact with Air Traffic Control.

ICAO procedures require that the decimal point is spoken as 'DECIMAL.' The FAA will honor such usage by military aircraft and all other aircraft required using ICAO procedures. At and above 18,000 feet MSL, FL 180, you must state the words 'Flight Level' followed by the flight level's separate digits.

The three digits of bearing, course, heading, or wind direction should always be magnetic. The word 'TRUE' must be added when it applies.

A description of 'speed' is followed by the word 'KNOTS.' The Air Traffic Controller may omit the word 'KNOTS' when using speed adjustment procedures.

The digits for Mach speed Number preceded by 'MACH.' The FAA uses Coordinated Universal Time - UTC or ZULU for all operations. The word 'LOCAL' or the time zone must denote when the local time is given during radio communications.

The factor of Time may be stated in minute's only two figures in radio communications, so no misunderstanding likely to occur. If you have reason to believe the receiver inoperative, you must remain outside or above the Class D surface area until the direction and flow of traffic have been determined. After that, you must advise the tower of your type of aircraft, position, altitude, intention to land, and request to be controlled with light signals.

When you approximately three to five miles from the airport, you must advise the tower and join the airport traffic pattern. You need to watch the Air Traffic Control and look for a light signal addressed to your aircraft from this point on. At night, you must acknowledge by blinking the landing or navigation lights. Before leaving the parking area, if you experiencing a radio failure, make every effort to have the equipment repaired. If you cannot have the malfunction repaired, you must call the tower by telephone and request authorization to depart without two-way radio communications. If the ATC tower grants authorization, it will give you departure information and a request to monitor the tower frequency or watch for light signals as it appropriated.

You may acknowledge the tower transmissions by moving the ailerons or rudder. At night, the acknowledgment is done by blinking the landing or the navigation lights. Some VOR is utilizing for voice channels for recording or broadcasting such as ATIS and HIWAS.

These services and other appropriate frequencies are listed in the A/FD.

# COMMUNICATION

At a controlled airport, you should adhere to the controller in an airport with an operating control tower that specifies the traffic pattern. At uncontrolled airports, traffic pattern altitudes and entry procedures may vary according to established local procedures.

Helicopter pilots should be aware of the standard traffic pattern and avoid it. Generally, helicopters make a lower altitude pattern opposite from the fixed-wing pattern and make their approaches to some point other than the runway in use by the fixed-wing traffic.

When a control tower is in operation, request the type of departure desired. In most cases, helicopter departures are made into the wind unless obstacles or traffic dictates otherwise.

At airports without an operating control tower, comply with the departure procedures established for that airport, if any. An accepted helicopter traffic pattern flown at 500 feet AGL and consists of right turns

That will keep the helicopter out of the flow of fixed-wing traffic.

A helicopter may take-off from a helipad into the wind with a turn to the right after 300 feet AGL or as needed in forced landing areas.

Depending on obstructions and forced landing areas, the final approach may need to accomplish from as high as 500 feet AGL. The landing area should always be insight, and the angle of approach should never be too high (indicating that the base leg too close) to the landing area or too low (indicating that the landing area too far away).

# Aeronautical Information Manual (AIM)

The Aeronautical Information Manual (AIM) is the best reference for learning good communication skills and phraseology. Because the FAA wrote it, the AIM is also the most authoritative source for IFR procedures. Unlike the federal aviation regulations, the AIM is not legally binding, but it is the most current and detailed source of FAA-recommended procedures. It's an art to using the right words to communicate with air traffic control (ATC). Effective aviation phraseology combines the transfer of complete and correct information.

If the ATC controller is busy with other aircraft and needs to issue you timely control instructions, they cannot do it until you release the microphone button.

This delay may affect the safety of the other helicopters. If the transmissions are too brief, the controller could ask you to provide more detail.

For a helicopter to proceed from one point to another, below 100 feet AGL and at airspeeds above 20 knots, use the following phraseology:

AIR TAXI:

TO - (what location, heliport, and active/inactive runway).

AVOID (aircraft/vehicles/personnel).

REMAIN AT OR BELOW (altitude).

CAUTION (wake turbulence or other reasons above).

LAND AND CONTACT TOWER,

HOLD FOR (reason-takeoff clearance, release, landing/taxiing aircraft, etc.).

CLEARED FOR TAKEOFF.

DEPARTURE FROM (requested location)

OWN RISK (additional instructions, as necessary).

HOLD SHORT OF (active runway, extended runway centerline, other).

CLEARED TO LAND.

ACKNOWLEDGE - Let me know you have received my message.

AFFIRMATIVE - Yes.

BLOCKED-Phraseology indicates that a radio transmission has been distorted or interrupted due to multiple simultaneous radio transmissions.

CLEARED FOR TAKEOFF - ATC authorization for an aircraft to depart.

CLEARED FOR THE OPTION - ATC authorization for an aircraft to make a touch and go, low approach, missed approach, stop and go, or complete-stop landing at the discretion of the helicopter pilot. It is normally used in training to evaluate a student's performance under changing situations.

CLEARED TO LAND - ATC authorization for an aircraft to land. It predicated on known traffic and known as the physical airport conditions.

CLOSED TRAFFIC - Successive operations involving takeoffs and landings (touch-and-go)] or low approaches where the aircraft does not exit the traffic pattern.

EXPEDITE - Used by ATC when prompt compliance is required to avoid the development of an imminent situation.

FLY HEADING (Degrees) - Informs the pilot of the heading. ATC may ask you to turn or continue to a specific compass direction to comply with the instructions. You may expect to turn in the shorter direction to a new heading unless otherwise instructed by ATC.

FUEL REMAINING - A phrase is used by either the helicopter pilots or the controllers when relating to fuel remaining on board until actual fuel exhaustion.

IMMEDIATELY - Used by ATC when such action compliance is required to avoid an imminent situation.

MAINTAIN-Concerning altitude/flight level, the term means to remain at the altitude/flight level specified.

MAKE SHORT APPROACH - Used by ATC to inform a helicopter pilot to alter his traffic pattern to make a short final approach.

MAYDAY - The international radiotelephony distress signal. When repeated three times, it indicates imminent and grave danger and that immediate assistance are requested.

MINIMUM FUEL - Indicates that an aircraft's fuel supply has reached a state where, upon reaching the destination, it can accept little or no delay. This is not an emergency but merely indicates an emergency is possible should any undue delay occur.

NEGATIVE - 'No,' or 'permission not granted,' or 'that not correct.'

NEGATIVE CONTACT - Used by pilots to inform ATC that the previously issued traffic is not in sight. It may follow by a request for a controller to assist in avoiding traffic. Used by pilots to inform ATC, they were unable to communicate on a particular frequency.

READ BACK - Repeat my message back to me.


RADAR CONTACT - Used by ATC to inform an aircraft identified on the radar display and radar flight following will be provided until the radar identification is terminated.

RADAR SERVICE TERMINATED - Used by ATC to inform a pilot that it will no longer provide any of the services that could be received while in radar contact.

REPORT - Used to instruct helicopter pilots to advise ATC of specified information; e.g., 'Report passing Hamilton VOR.'

SAY AGAIN - Used to request a repeat of the last transmission. Usually, specified transmission or portion is not understood or received; e.g., 'Say again all after.'

SAY ALTITUDE - Used by ATC to ascertain an aircraft's specific altitude/flight level. When the aircraft climbing or descending, you should state the indicated altitude rounded to the nearest 100 feet.

SAY HEADING - Used by ATC to request an aircraft heading. You should state the actual heading of the aircraft.

SPEAK SLOWER - Used in verbal communications as a request to reduce the speech rate.

SQUAWK (Mode, Code, and Function) - Activate specific modes/ codes/functions on the aircraft transponder, e.g., 'Squawk two-one-zero-five.' Squawk does not mean that the pilot should press the transponder's IDENT button.

STAND BY - This means the controller or the pilot must pause for a few seconds, usually to attend to other duties of higher priority. Besides, it means to wait as 'stand by for clearance.' The caller should reestablish contact if a delay lengthy. 'Stand by' is not an approval or denial.

TAXI INTO POSITION AND HOLD - Used by ATC to inform a pilot to taxi onto the departure runway in takeoff position and hold. It not authorization for takeoff. It is used when takeoff clearance cannot immediately issue because of traffic or other reasons.

THAT CORRECT - the understanding you have it right.

TRAFFIC - A term used by ATC to refer to one or more aircraft.

TRAFFIC IN SIGHT - Used by helicopter pilots to inform a controller that previously issued traffic insight.

UNABLE - Indicates the inability to comply with a specific instruction, request, or clearance.

VERIFY - Request confirmation of information; 'verify assigned altitude.'

# Helicopter pilot's Alphabet

Pilots use code words to communicate letters of the alphabet. Pilots must use the alphabet, so those essential letter combinations can easily be understood by whoever receives the voice messages.

| A | Alfa | AL-FAH |
|---|------|--------|
| B | Bravo | BRAH-VOH |
| C | Charlie | CHAR-LEE |
| D | Delta | DELL-TAH |
| E | Echo | ECK-OH |
| F | Foxtrot | FOKS-TROT |
| G | GOLF | Golf |
| H | Hotel | HOH-TEL |
| I | India | IN-DEE-AH |
| J | Juliet | JEW-LEE-ETT |
| K | Kilo | KEY-LOH |
| L | Lima | LEE-MAH |
| M | Mike | MIKE |
| N | November | NO-VEM-BER |
| O | Oscar | OSS-CAR |
| P | Papa | PAH-PAH |
| Q | Quebec | KWEH-BECK |
| R | Romeo | ROW-ME-OH |
| S | Sierra | SEE-YE-RAH |
| T | Tango | TANG-GO |

| U | Uniform | THE HELICOPTER PILOT-NEE-FORM |
|---|---------|-------------------------------|
| V | Victor | VIK-TAH |
| W | Whiskey | WISS-KEY |
| X | X-ray | EKS-RAY |
| Y | Yankee | YANG-KEY |
| z | Zulu | ZOO-LOO |
| | | |
| 1 | One | WUN |
| 2 | Two | TOO |
| 3 | Three | TREE |
| 4 | Four | FOW-ER |
| 5 | Five | FIFE |
| 6 | Six | SIX |
| 7 | Seven | SEV-EN |
| 8 | Eight | AIT |
| 9 | Niner | NIN-ER |
| 0 | Zero | ZEE-RO |

# AIRSPACE DETAILS

Controlled airspace consists of those areas where some or all aircraft may be under Air Traffic Control within those airspaces.

• Class A Airspace is depicted as an open area (white) on the charts. It consists of airspaces from 18,000 MSL to FL600.

• Class B Airspace is a screened blue area with a solid line encompassing the area.

• Class C Airspace a screened blue area with a dashed line encompassing the area.

• Class B and Class C Airspaces are controlled airspace extending upward from the surface or designated floor to specified altitudes.

• Class D Airspace (airports with an operating control tower) is an open area (white) with a "D" enclosed within a box following by the airport name.

• Class E Airspace is an open area (white) on the En-route Low Charts of airspaces below 18,000 MSL.

• Airports within the airspace, which fixed-wing special VFR flights, are prohibited, shown as NO SVFR.

The boundaries of the ATCs centers are shown in their entirety using the symbol below. Center names are shown adjacent and parallel to the boundary line. Their frequencies are shown in boxes outlined by the same symbol.

# Airspace

Other airspace areas are a general term referring to the majority of the remaining airspace. It includes:

• Airport Advisory Areas

• Military Training Routes (MTR)

• Temporary Flight Restrictions

• Parachute Jump Areas

• Published VFR Routes

• Terminal Radar Service Areas

• National Security Areas

Mode C Required Airspace (from the surface to 10,000' MSL) within a 30 NM radius of the primary airport(s) for which Class B airspace is designated, drawn on the En-route Low Altitude Charts. Mode C is also marked within 10 NM of all airports.

Mode C is required within the limits of class C airspace up to 10,000 ft. MSL.

Controlled airspace is a generic term that covers the different classifications of airspace. It defines the dimensions with air traffic control services provided under the airspace classification.

The two categories of airspace are regulatory and none regulatory. There are four types: controlled, uncontrolled, special use, and other various classes of the airspace. Airspaces are named by alphabetical letters. Even if you are not allowed into Classes A, B, C, D, Restricted, or Prohibited Areas, you are allowed into nearly all Class E and all of Class G airspace.

# ICAO

ICAO a regulatory body, not a direct ATC service. The international ATC delegated an identification number to nations who accept responsibility for providing ATC services. As such, they are the guiding body when flying in international airspace. Most domestic and foreign governments are requiring their aircraft pilots to abide. ICAO has divided the airspaces in the world into Flight Information Regions. These regions identifying which country controls the airspace and determines which of the procedures to use. Normally, one major ATC facility is identified with each FIR. These facilities are called Area Control Centers (ACC). They equivalent to ARTCC are in the U.S.

# Class A Airspace:

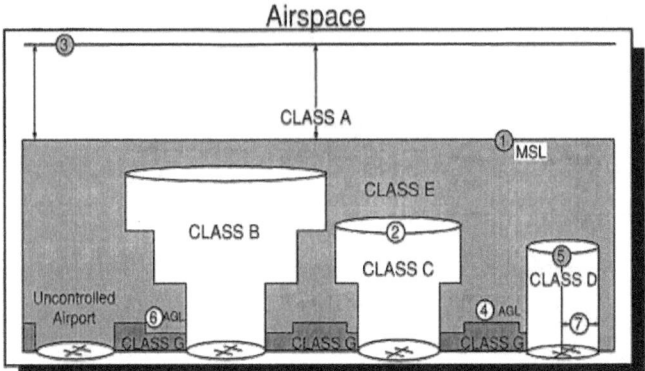

The use of ATC service is mandatory in this airspace. Class A begins at 18,000 feet MSL and extends upward to 60,000 feet MSL. Class A airspace does not show on the sectional map, but it covers the entire nation, so we need to remember that the lower limit of Class A is 18,000 MSL. All altitudes at 18,000 feet MSL and above, in Class A airspace, referred in thousands of feet as 'Fight Levels,' and it's abbreviated as - FL.

You could say - flight level two zero, zero, or FL 200 = 20,000 feet. The United States government has extended Class A airspace out to twelve miles from the coast of the contiguous 48 states and Alaska.

# Class B Airspace:

The topmost of Class B airspace extends to a radius of 15 nautical miles around the airport tower. The B airspace border with the country's busiest airports, usually its verve up to 10,000 feet. However, there no universal set of Class B dimensions since the flow of traffic is determined the exact architecture of each Class B area. On a sectional map, the horizontal Class B airspace-limits, outlined in a solid circular blue line. They may be concaved or stretched in certain places due to topography and air-traffic routes. The top and bottom of each airspace are displayed in what looks like a fraction; for example, 90/40. That means this particular layer of airspace lies between 9,000 and 4,000 feet MSL. This is what the FAA would call the "congested" area. Class B airspace is extremely dangerous to the lives of many since many passenger airlines are carrying thousands of passengers. Pilots' violation of Class B or Class C will almost certainly bring strong penalties. The surface area of a city in Class B airspace is colored yellow, like all cities on this sectional.

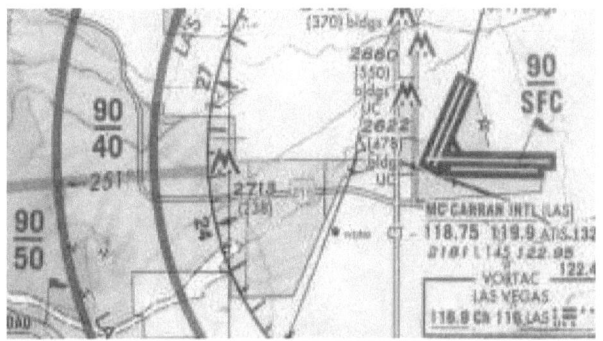

# Class C Airspace:

On your sectional, solid magenta lines show horizontal Class C limits. in layout, It's similar to Class B but drawn in magenta, not in blue.

The upper and lower limits are applied to Class B. Pilots may fly under or over Class C airspace, but never into the airspace.

The cities under class C airspace are mid-sized cities. The towers in these fields are equipped with radar -- something that smaller controlled fields (Class D airspace) do not have.

# Class D airspace

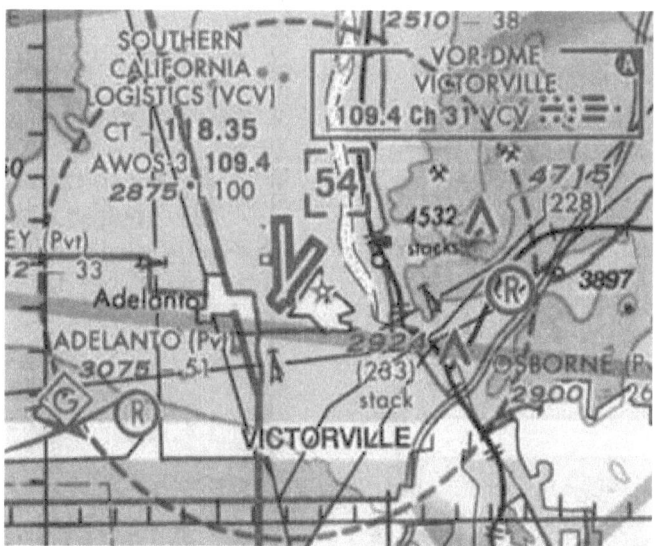

Class D airspace is designated for small city airports with control towers and commercial IFR traffic. They often have much of the general aviation activity and Pilot training.

The field silhouettes itself (not the city) is visualized in blue color with a dashed blue circle around it. All runways are drawn to show their direction in terms of the compass, and that runway length is given.

The Class D airspace ceiling extends upward to 2,500 feet AGL over the airport surface, but the exact upper limit is shown with a number inside a dashed box outline.

The number '54' means that the upper limit 5,400' MSL. Some Class D fields have little extensions, which look like cogs in a wheel. The entire airspace may look like a keyhole with one or more extensions out of the five-mile circle.

The Class D airspace tower may not have radar.

# Class E airspace

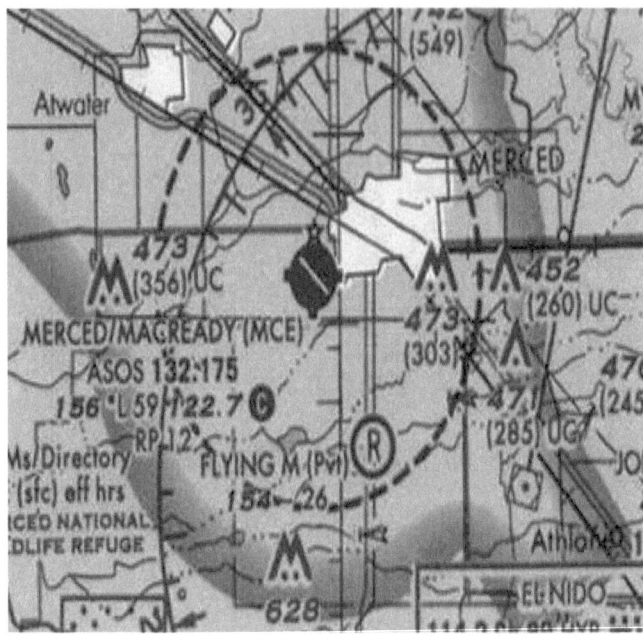

Class E airspace is controlled airspace, not Class A, B, C, or D or G airspace. The filler fills in under Class A, which means; it's below 18,000', and between Classes B, C, and D, and over the top of Class G. If we ignore the upper cover of Class A airspace for a minute, it is safe to say that there is a lot more of E Class than all the other kinds of airspaces combined. Class E usually has four limits. The Surface, 700' AGL, 1200' AGL, and 14,500'.

In most of the country, the Class E lower limit is 1200' AGL.

When it drops to 700' AGL, a broad magenta line with a fuzzy side will show it. The fuzzy side is where the floor of Class E 700' AGL.

The Class E boundaries are displayed in dashed-line magenta color. You will need a prior authorization from the ATC to enter it.

Class E Area looks much like Class D but only in a dashed magenta, not dashed blue, and no upper altitude number within a box-like in Class D. A number in blue on the side of the line indicates that this the floor of Class E.

# Class G airspace

Class G airspace is a mantle of low-lying airspace beginning at the surface. Class G airspace is entirely uncontrolled. This low-lying blanket of uncontrolled airspace only ends when it meets Class B, C, D, or E airspace. It covers the entire country. In very remote areas, the upper limit is at 14,500' MSL.

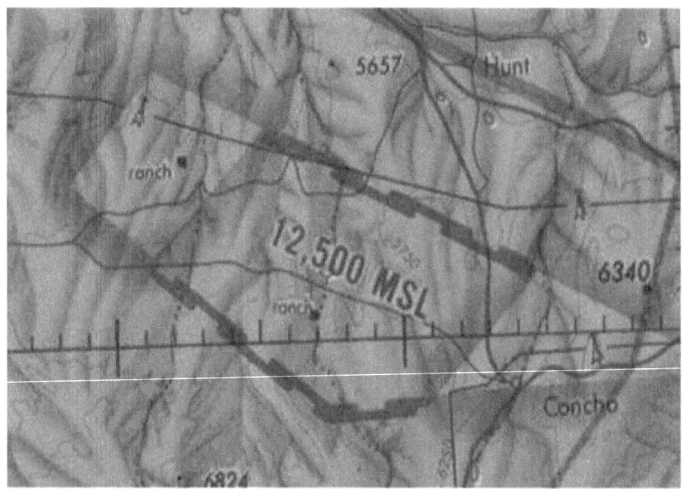

The floor of Class E is on top of Class G. The depth of Class G is 700', and sometimes its 1200'.

Class B, C, D, or E extends to the surface, and there no surface to Class G.

In the early days of aviation, all airspaces were uncontrolled, what we now call the Class G airspace. There were only a few planes at the time, and none had the instruments necessary to fly in the clouds. Even the traffic density the lowest. And the aircraft flew slower.

# Special use Airspace:

When planning a flight, you are expected to avoid areas called Special Use Airspace (SUA). There are several types of SUA in the United States, including a Restricted, Warning, Prohibited, Alert, and Military Operations Area (MOA).

# Prohibited area

Victor Airways are similar to highways but are in the sky. Many aircraft follow these routes. The routes connect by radio navigation beacons "very high-frequency omnidirectional range" or VOR stations that radiate signals in all directions. These stations are located at or near the airports. The North-south Victor Airways have odd numbers while the east-west airways have even numbers. Both IFR and VFR aircraft use these federal or Victor Airways.

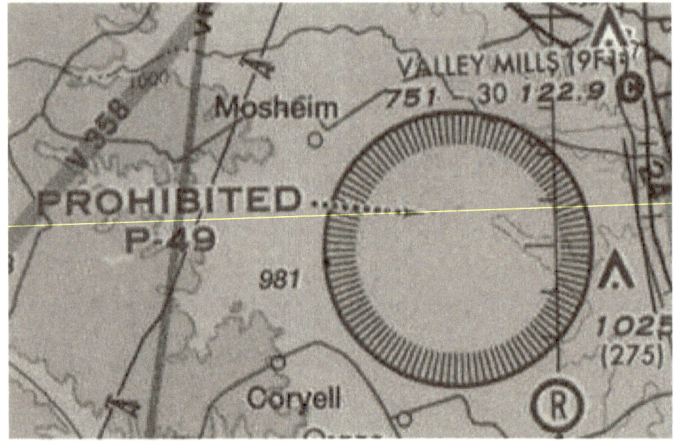

This airspace is set aside for a Victor Airway. Its eight miles wide with a floor at 1200 AGL. The airspace extends up to FL 180. Victor Airways are drawn on your sectional map by a faint blue line printed the V designation along with the airway number.

# Military Training Routes

Military Training Routes or MTRs plotted as thin, are light gray lines on the sectionals. Each of them has its identifier, and that identifier has two parts.

"VR"- means that Pilots flying the training routes will fly under visual flight rules.

"IR" - means the Pilots will be flying under instrument flight rules.

The second identifier is either a three or four-digit number. Four digits mean the route will be flown at or below 1500 feet AGL A three-digit number means the route will be flown both below and above 1500 feet AGL.

VR-1260 means - a training route is flown under VFR at a low level.

IR-141 - means a training route is flown under IFR at a low level.

# Warning Area

This hazardous area lies over international waters is beyond the three-mile coastal limit. Military Operations Area (MOA), These large areas of the country, are shown on your sectional enclosed by the hash-marks magenta line and sharp outer edge.

Military operations, such as exercises, training, come and go. Permission is not required to fly in an MOA, and the pilot may determine the hours of activity.

Alert Area - Bordered the same as a Prohibited Area, the identifier is not with an R or P but with an A. You allowed flying in the Alert Area without prior authorization. An Alert Area may involve high aviation traffic, special air operations, or frequent student training.

The area is marked with a blue border and with a word or two of explanation.

Wildlife and recreational areas -- These sensitive areas indicated on the sectional with lines of blue dots. The minimum altitude for fly-over 2,000 AGL. A violation here could result in penalties. Landing in such areas may be in an emergency only.

A- stand for any airspace Above FL 180 and up to FL 600.

B - stands for Big Time or Big City, complex Class B airspace.

C - stands for Cities of moderate size with air traffic control.

D - stands for Dime Sized or Diminutive cities with controlled traffic. E - stands for Elemental, Everywhere, or Universal airspace. G - stands for Go-For-It, uncontrolled airspace.

V - stands for Victor Airway

MOA - stands for Military Operating Area.

# Restricted areas

The Restricted areas include airspaces characterized by an area on the surface of the Earth in which the flight of aircraft, although not completely prohibited, is restricted. The activities in these areas should be restricted because of their nature or restrictions on aircraft operation that is not part of these activities. Restricted areas indicate an unusual hazard, often invisible to aircraft, such as artillery fire, air rifles, or guided missiles. Unauthorized entry into the restricted areas can be hazardous for the aircraft. Restricted areas are published in the Federal Register and form 14 CFR Part 73.

ATC facilities use the following procedures when aircraft with IFR clearance (including those approved by the ATC to guarantee visibility clearance) fly over a route in a controlled or restricted area.

If the restricted area is not active or surrendered to the FAA, the air traffic control allows aircraft to operate in the restricted area without special authorization.

Suppose the restricted active area has not been surrendered to the control authority (FAA). In that case, the air traffic controller will issue a permit to ensure that the aircraft bypasses the restricted area unless it's either on an approved flight reservation mission or has received his own approval to operate in that airspace informs the controller.

# Special Use Airspace

Special Purpose Airspace or Special Area of Operation (SAO) is the name of the airspace. Certain activities are restricted or in which the restrictions can be imposed on that do not include these activities. Certain areas of the airspace design for special purposes and can be restricted in mixed-use. The special-purpose airspace is shown on the instrument cards. It contains the name or number of the area, the actual altitude, operating time and weather conditions, the control point, and the card sign location. This information is available on the National Aeronautical Charting Group (NACG) route maps on one of the end gauges. General-purpose airspace generally includes:

Forbidden areas

Closed areas

warning areas

Military Areas (MOA)

alarm zones

# Controlled Fire Areas (CFA).

A **controlled firing area (CFA)** airspace is designated to contain activities that would be hazardous to a nonparticipating aircraft if not conducted in a controlled environment.

The Federal Aviation Administration has established a Controlled Firing Area (CFA). The CFA is activated only during the times needed to support specific training and the altitude necessary to contain each day's activities. It will be deactivated immediately upon completion of training each day.

Controlled Firing Areas are one of the six special use airspaces.

The special use airspaces listed below:

Prohibited areas (regulatory)

Restricted areas (regulatory)

Warning areas

Military operation areas (MOAs)

Alert areas

Controlled Firing Areas (CFAs)

# WEIGHT AND BALANCE

Total weight is a fixed value: the weight of the helicopter, fuel, and occupants. The rotor blades must generate enough lift to overcome the total weight and lift the helicopter off the ground vertically.

The manufacturer provides the aircraft operator with the empty weight of the aircraft and the location of its empty-weight center of gravity (EWCG) at the time the certified aircraft leaves the factory.

Light helicopters are those generally considered below 12,000 pounds [5,000 kg] maximum gross weight. Medium helicopters are generally considered about 14,000 pounds to 45,000 lbs. (6,000 kg to 20,000 kg).

Aerodynamic loads can also influence the weight of the helicopter. The load factor is the actual load on the rotor blades at any time, divided by the normal load or gross weight (weight of the helicopter and its contents).

Regardless of how much weight one can carry or the engine power they may have, they are all susceptible to aerodynamic overloading.

If you attempt to push the performance, the consequence can be fatal. Aerodynamic forces affect every helicopter movement, whether it increases the Collective or steep the bank angle.

As a pilot, you are responsible for knowing the maximum allowable weight of the aircraft and its CG limits. It allows you to determine the preflight inspection if the aircraft overloaded and whether the CG within the allowable limits.

Anticipating results from a particular maneuver or adjustment of flight control, not a good helicopter piloting technique. Instead, helicopter pilots need to understand the helicopter's capabilities under all circumstances and plan not to exceed the flight envelope for any situation.

Weight a major factor in helicopter construction and operation, and it demands respect from all Helicopter pilots and particular diligence by all A&P mechanics and repair people. Too much weight reduces an aircraft's efficiency and the safety margin available if an emergency condition should arise.

\* \* \* \* \*

When a helicopter is properly loaded, the CG is within the allowable CG range, and will it stay within the allowable range throughout the flight, including all loading configurations.

Ensure that the total weight does not exceed the maximum allowable gross weight. Use CG or moment information from loading charts, tables, or graphs in the RFM. Weight and balance computation are only as accurate as the information provided.

Ask passengers what they weigh and add a few pounds to account for the additional weight of clothing, especially during the winter. Baggage should be weighed on a scale. If a scale is not available, compute personal loading values according to each individual estimate.

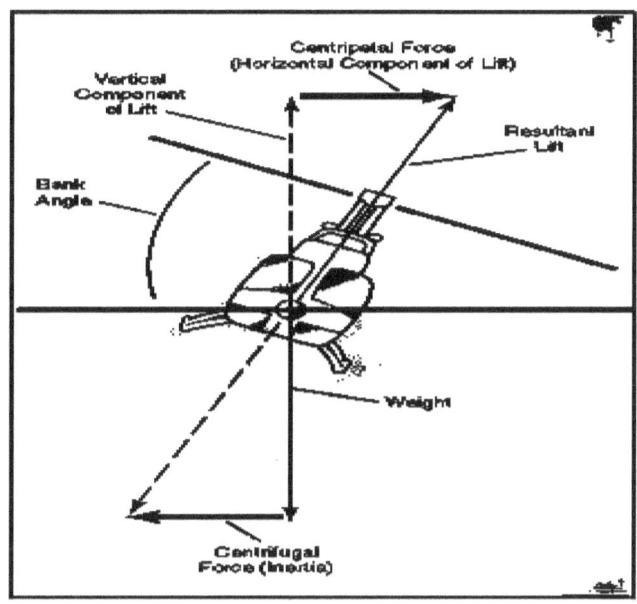

Following Title 14 of the Code of Federal Regulations (14 CFR) part 23, a normal category helicopter structure must be strong enough to sustain a load factor of 3.8 times its weight.

* * * * *

Every pound of weight added to an aircraft requires that the structure is strong enough to support an additional 3.8 pounds.

The center of gravity (CG) is the point of balance along the helicopter's longitudinal axis. By multiplying each helicopter's component's weight by its arm (distance from an arbitrary reference point, known as the reference datum), you could determine the component's moment.

The CG of the helicopter is the sum of all the moments divided by the total weight. The 6 lbs. per gallon, gasoline became the standard weight in aviation(AVGAS), and so the 25 gallons of AVGAS equal to 150 lbs. (25 x 6 = 150).

The FAA requires a current and accurate empty weight and empty weight center of gravity known for an aircraft. The weight and balance report must result in the aircraft at the time of operation.

Weight and balance for large helicopters almost identical to small helicopters but on a much larger scale.

If a technician could weigh a small helicopter and calculate its empty weight center of gravity, that same technician could do it for a large helicopter. The jacks and scales will larger, and it may take more personnel to handle the equipment, but the concepts and processes the same.

When determining if a helicopter within the weight limits, consider the basic helicopter's weight, the crew, passengers, cargo, and fuel.

When operating in a confined area, planning must ensure that the helicopter can lift the weight during flight phases. The weight may be acceptable during the early morning hours. Still, as the density altitude increases during the day, the maximum allowable weight may have to reduce to keep the helicopter within its capability.

Most helicopters have an internal maximum gross weight, which refers to the weight within the helicopter structure and an external maximum gross weight.

On small helicopters, the center gravity is located at a specific number of inches from the datum. The center of gravity is identified in the same way. The width of the wing of an aircraft is known as the chord.

The designers determine the center of gravity (CG) 's the ideal location, and the maximum deviation allowed has already been calculated.

You have the responsibility on every flight to know the aircraft's CG limits. Too much weight will reduce the efficiency of an aircraft and the safety margin available in an emergency. When increasing the weight, the rotors should produce additional lift, and the structure supports the additional static loads and the dynamic loads imposed by flight maneuvers.

Some large cargo helicopters may have several attachment points for sling load or winch operations. These helicopters can carry a tremendous amount of weight when the attachment point directly under the CG of the aircraft.

To accurately predict the helicopter performance, it's necessary to guarantee the helicopter's structural integrity and weight limitations.

You should never intentionally exceed the certified load limits for that helicopter. Operating below a minimum weight could adversely affect the handling characteristics of the helicopter.

In some helicopters, a helicopter pilot needs to use a large amount of forwarding Cyclic to maintain a hover. When operating at or below the minimum weight of the helicopter, the additional weight also improves the auto rotational characteristics since the auto rotational descent was established sooner.

Operating below minimum weight could prevent achieving the desirable rotor revolutions per minute during the autorotation. Operating above a maximum weight could result in structural deformation or failure during the flight if encountering too much load factors, strong wind gusts, or turbulence.

Anything that adversely affects takeoff climb, hovering and landing performance may require off-loading fuel, passengers, or baggage to some weight less than the published maximum.

Severe uncoordinated maneuvers or flight turbulence could impose a dynamic load on the structure, great enough to cause failure.

Some manufacturers specify this range as measured in percentage of the mean aerodynamic chord (MAC), the leading edge located at a specified distance from the datum.

The datum location may be anywhere the manufacturer chooses; it's often the leading edge of the wing or some specific distance from an easily identified location. A moment is a force that tries to cause rotation, and the arm's product, measured in inches, and the weight, measured in pounds. Moments are generally expressed in pound-inches (lb. In) and may either be positive or negative. Positive moments cause an aircraft to nose-up, while the negative moments cause it to nose down.

A lever is balanced when the weight on one side of the fulcrum multiplied by its arm, is equal to the weight on the opposite side multiplied that balance is the algebraic sum of the moments equal to zero.

When an item in the equipment list is added or removed from the aircraft, its weight is used to update the record's new weight and balance.

The weight and balance report have comprised an equipment list by showing weights and moment arms of all required and optional items of equipment comprised in the certificated empty weight.

When an aircraft has undergone an extensive repair or major alteration, it should re-weigh, and a new weight and balance record started.

The Empty weight does not include the engine oil. You need to drain the oil before weighing the aircraft or subtract it from the scale readings to determine the empty weight.

If it is impractical to drain the oil, you could fill the reservoir to the specified level and compute the oil weight at 7.5 pounds per gallon. Then its weight and moment subtracted from the weight and moment of the aircraft as weighed.

The Moment factor is resolved by multiplying the weight of each component by its arm. To get the moment index, you should divide the moment by reducing factors such as 100 or 1,000. The loading graph will provide you with the moment index for each component to avoid mathematical calculations.

The CG envelope uses moment indexes rather than arms and moments. If you draw lines of weight and CG and cross within this envelope, the helicopter is properly loaded.

Some large transport helicopters have an on-board aircraft weighing system (OBAWS) that, when the aircraft is on the ground, gives the flight crew a continuous indication of the aircraft's total weight and the CG location in % MAC.

The system consists of strain-sensing transducers in each main wheel and the nose wheel axle, weight and balance computer, and indicators that show the gross weight. The CG location is in percent of MAC and an indicator of the aircraft's ground altitude. This is a method for calculating and documenting aircraft weight and the balance before taxiing, to ensure the aircraft will remain within all required weight and balance limitations throughout the flight.

The moment is a force that causes or tries to cause an object to rotate. Moment Index is the moment (weight times arm) divided by a reduction factor such as 100 or 1,000 to make the number smaller and reduce the chance of mathematical errors in computing the center of gravity.

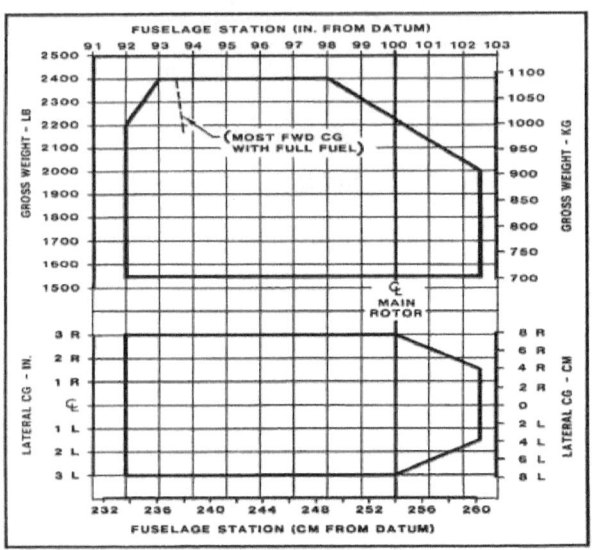

**CENTER OF GRAVITY LIMITS**

Helicopter performance is not only affected by the gross weight, but also by the position of that weight. It's essential to load the aircraft within the allowable CG range, specified in the RFM's, weight and balance limitations. You should try and balance a helicopter is perfectly so that the fuselage will remain horizontal in hovering flight, with no cyclic pitch control needed, except for wind correction. Since the fuselage acts as a pendulum that's suspended from the rotor. Changing the CG changes the angle at which the aircraft hangs from the rotor.

When the CG is directly under the rotor mast, the helicopter hangs horizontally; if the CG too far forward of the mast, the helicopter hangs with its nose tilted down; if the CG too far aft of the mast, the nose tilts up.

It is generally unnecessary for smaller helicopters to determine the lateral CG for normal flight instruction and passenger flights. It happens because helicopter cabins are relatively narrow, and most optional equipment is near the centerline.

Suppose they're an unusual situation that could affect the lateral CG, such as a heavy helicopter pilot and a complete fuel load on one side of the helicopter. In that case, its position should be checked against the CG envelope.

If carrying an external load in a position that requires large lateral Cyclic control displacement, so to maintain leveled flight, it could limit the fore and aft cyclic effectiveness.

Manufacturers generally account for the known lateral CG displacements by locating an external attachment point opposite the lateral imbalance.

# NAVIGATION AND FACILITIES

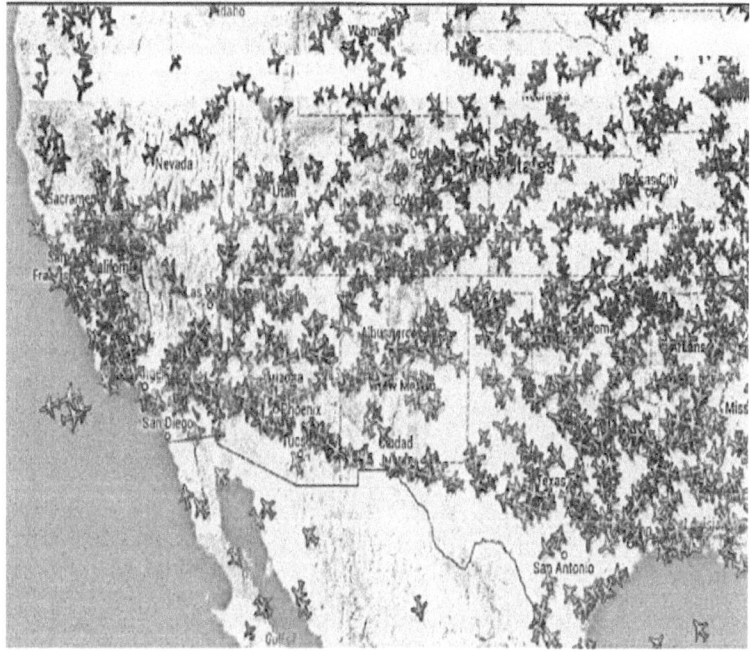

Air navigation is the process of piloting an aircraft from one place to another while monitoring its position as the flight progresses. An aeronautical chart is the road-map for a pilot flying under VFR.

In addition to the information automatically comprised in an AWOS report, you can manually add the information. The remarks are limited to thunderstorms (type and intensity) and obstructions to vision when the visibility is 3 SM or less. Augmentation identifies the observation as observer weather.

# Latitude and Longitude

The equator is an imaginary circle from the poles of the Earth. Circles parallel to the equator (lines running east and west) parallels of latitude. They measured the degrees of latitude north (N) or south (S) of the equator. The angular distance from the equator to the pole is one-fourth of a circle or 90°. Meridians of longitude are drawn from the North Pole to the South Pole and right angles to the Equator. The 'Prime Meridian,' which passes through Greenwich, England, is used as zero lines to measure from, and it's displayed in degrees, east (E), west (W) up to 180°.

The United States is between 67° and 125° W longitude. This way, any specific geographical point could refer to the location by its longitude and latitude.

# TIME ZONES

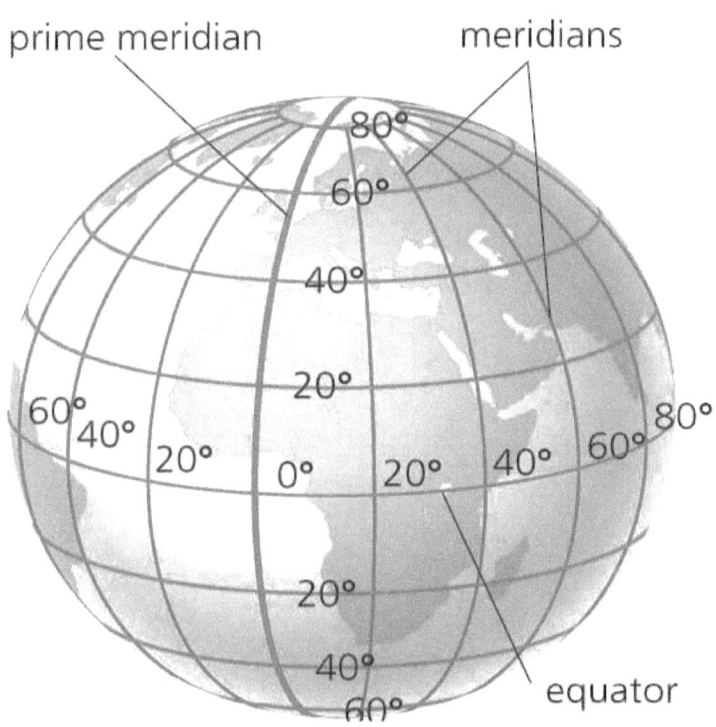

prime meridian    meridians

80°
60°
40°
20°
60° 40°
20°
0°    20°    40°    60° 80°
20°
40°
60°

equator

The meridians are also useful for designating time zones. A day defined as the time requires the Earth to make one complete rotation of 360°. For a day of 24 hours, the Earth revolves at the rate of 15° an hour.

Noon is when the sun is positioned directly above a meridian. To the west of that meridian, its morning, and to the east, its afternoon. This is the standard practice to establish a time zone for each 15° of longitude. The difference between each zone is exactly 1 hour. In the United States, there four time zones.

The time zones are Eastern (75°), Central (90°), Mountain (105°), and Pacific (120°). When the sun directly above the 90th meridian, its noon - Central Standard Time. These time zones differences are taken into account during long flights in eastward direction - especially if they need to be completed before darkness. An hour is lost when flying eastward from one time zone to another, or perhaps even when flying from the western edge to the same time zone's eastern edge.

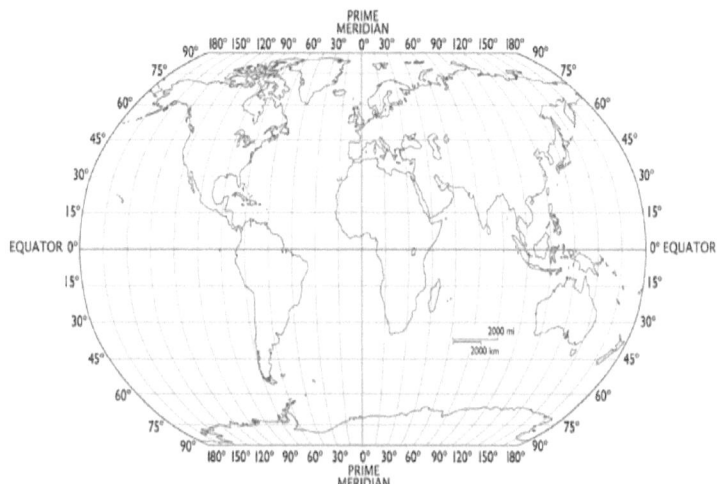

In most aviation operations, time is a 24-hour clock. In ATC instructions, weather reports and broadcasts, and estimated arrival times are all based on this system. For example, 9 a.m. expressed as - 0900, and 10p.m. - 2200.

Because a pilot may cross several time zones during a flight, a standard time system must apply

It is called Universal Coordinated Time (UTC) and often is referred to as Zulu time. UTC is the time at the 0° line of longitude, which passes through Greenwich, England.

All-time- zones around the world are based on this reference. To convert to this time, you should do the following:

Eastern Standard Time.......... Add 5 hours

Central Standard Time.......... Add 6 hours

Mountain Standard Time...... Add 7 hours

Pacific Standard Time.......... Add 8 hours

During Daylight Saving Time, you should subtract one hour from the calculated times.

Using the meridians, you can measure the direction in degrees from one point to another, in a clockwise direction from the true north. To indicate a course to be followed in flight, draw a line on the map chart from the departure point to the destination, and then measure the angle. This line forms with a meridian – the direction is articulated in degrees.

On a chart map, magnetic meridians are shown in red while the longitude lines and the latitude are blue. From these lines of variation (magnetic meridians), you can determine the effect of local magnetic variations on a magnetic compass.

# Distance measuring equipment

Distance measuring equipment (DME) is a transponder-based radio navigation technology that measures range angle distance by timing the propagation delay of the VHF or UHF radio signals.

DME is similar to secondary radar, except in reverse. DME is functionally identical to the distance-measuring component of TACAN. Pilots use the DME to determine their distance from a land-based transponder by sending and receiving pulses pairs – two pulses of fixed duration and separation. Most VORs are co-located at the ground stations.

A low-power DME is co-located with the ILS Localizer antenna. It has a list of 1 kW peak power pulse output.

The pilots use in En-route DME ground transponder system for navigation. It provides an accurate distance to the touchdown function and is similar to the ILS Marker Beacons.

The DME system is composed of a UHF transmitter/receiver in the aircraft and a UHF receiver on the ground.

The DME receiver in the aircraft will search for a replied pulse-pairs, with the correct interval pattern.

The aircraft interrogator locks on to the DME ground station once it recognizes a particular reply pulse sequence with the same spacing as the original interrogation sequence. Once the receiver is locked on, it has a narrower window to look for the echoes and retain the lock.

An encoding transponder and two-way radio are required to operate within Class C airspace. A Mode C transponder is also required. Two-way communications are also required, but the DME is not required.

If the distance measuring equipment (DME) becomes inoperative En-route, you must notify ATC of that failure as soon as it occurs.

It measures nautical miles with a slant-range error on a distance information's display of the DME equipment. The greatest difference between the displaying distance and ground distance will occur at high altitudes and close to the VORTAC.

At low altitudes, close to the VORTAC, the slant-range error is less than at higher altitudes and closer to the VORTAC. The slant-range error is at its smallest at low altitudes far from the VORTAC.

The aircraft interrogator locks on to the DME ground station once it recognizes a particular reply pulse sequence with the same spacing as the original interrogation sequence. Once the receiver is locked on, it has a narrower window to look for the echoes and retain the lock.

# EMERGENCIES & HAZARD

# Coriolis Illusion

Studies show that when a helicopter pilot spends on visual tasks inside the cabin, it should represent no more than 1/4 to 1/3 of the scan time outside or 4 to 5 seconds on the instrument panel for every 16 seconds outside. An abrupt head movement while making a prolonged constant-rate turn could produce a strong rotation or movement sensation in an entirely different axis. The phenomenon is known as the Coriolis Illusion.

In the dark, a stationary light will appear to move about when stared at for many seconds. This illusion is known as Auto Kinesis.

Hypoxia appears as a result of too little oxygen reaching the brain. An excessive carbon dioxide in the bloodstream causes hyperventilation. It is the result of insufficient oxygen to the brain. The low pressure of oxygen causes hypoxia.

The percentage of nitrogen, carbon dioxide, and oxygen in the atmosphere remains constant with altitude changes, but they're less pressure as you increase altitude. The percentage of nitrogen, carbon dioxide, and oxygen in the atmosphere remains constant with altitude changes.

A rapid acceleration during takeoff could create the illusion of being in a nose-up attitude. It's known as a Somatogravic Illusion. The illusion is caused by an abrupt change when you climb to straight-and-level flight, which creates an illusion of tumbling backward. Auto Kinesis refers to a stationary light that appears to move about when staring at it for many seconds in the dark.

Hyperventilation symptoms include dizziness, tingling of the extremities, a sensation of body heat, rapid heart rate, blurring of vision, muscle spasm, and finally, unconsciousness. Any aircraft that appears to have no relative motion and stays in one scanned quadrant will likely be in a Coriolis illusion. It's likely to on a collision course. You should take evasive action.

In the absence of ground features, as when landing over water, in dark areas, wide and featureless terrain are made by snow, it could create an illusion of being at a higher altitude than it really is.

You are most likely to hyperventilate when under stress or at high altitudes. A slow breathing rate is a cure for hyperventilation. Insufficient oxygen is the cause of hypoxia and not hyperventilation. Carbon monoxide poisoning produces the same symptoms as hypoxia, which includes dizziness.

Rapid breathing could result in hyperventilation, but it is not a symptom of carbon monoxide poisoning. Tingling in the extremities(no pain or cramping) is only one symptom of hyperventilation.

An airplane on the horizon without movement may travel in the same direction as you. Rain on the windscreen could create an illusion of it being at a higher altitude than you are.

# Helicopter pilot error

There are several classic behavioral traps, which tumbled pilots.

These dangerous behavior patterns include peer pressure, mindset, get-there-itis, duck-under syndrome, continuing visual flight rules into instrument conditions, loss of position or situation awareness, operating without adequate fuel reserves, descent below the minimum En-route altitude, flying outside the envelope, neglect of flight planning, preflight inspections, checklists, etc.

It certainly is desirable to prevent as many errors as possible. Still, since you cannot prevent them all, you should address it by training the detection and recovery of errors.

Experienced helicopter pilots are likely to make automatic decisions. But there are tendencies or operational pitfalls that come with the development of pilot experience. These are the pilots' classic behavioral traps. More experienced helicopter pilots (as a rule) try to complete a flight as planned, please the passengers, and meet their schedules. The desire to meet these goals could harm the flight's safety and contribute to an unrealistic assessment of piloting skills.

Problem detection is the first step in the decision-making process. It begins with recognizing that a change occurred or that an expected changes that did not occur. A problem is perceived first by the senses, and then it is distinguished and determined through insight and experience. These same abilities and objective analysis of all available information are used to determine the problem's nature and severity.

One critical error made during the decision-making process is incorrectly detecting the problem. Therefore, incorrectly detecting a problem, to begin with, is an error that is critical during a decision-making process.

Effective monitoring and crosschecking could be the last defense line that prevents an accident because detecting an error or unsafe situation may break the chain of events leading to an accident. This monitoring function is always essential, and particularly during approach and landing.

Controlled Flight into Terrain (CFIT) accidents are the most common.

Pilots should consider error management, prevention, detection, and recovery when evaluating the prevention of errors.

# low rotor speed

During flight training, most helicopter pilots are trained to react to low-rotor RPM situations.

Robinson helicopter pilots get extra training every two years (per SFAR 73) because of the unusually high number of low-RPM accidents in the early Robinson helicopters. The new Robinsons have correlators and governors to help the pilot maintain proper speed, this special training and endorsement are still required.

Treat low rotor speed as an emergency procedure until it becomes your second nature.

To restore speed, immediately roll the throttle on, lower the collective, and apply aft cyclic in forwarding flight.

As a result, when you hear that low rotor speed warning horn — which is directly related to the deterioration of life-giving speed — react quickly to recover the lost speed. That means increasing the throttle (to add engine power) and lowering the collective (to reduce drag force caused by the rotor blades). Pulling back on the cyclic, when moving forward, can also help recover lost speed by transferring the energy in the forward speed to the rotor speed (which why speed increase during a cyclic flare in an autorotation). Rolling on the throttle is most important, but it doesn't always resolve the problem.

Lowering the collective frequently may help to hold that collective. But lowering the collective will cause you to descend faster than you already are. If you can bring the helicopter into a hover.

Even if the engine's speed will drop to 0 when you only a few feet off the ground, you're not going to die.

You'll drop like a brick —for a few feet. The skids will spread a little. Remember, in an R22 or R44, the horn sounds at 97% RPM.

You could still remain in flight with the RPM all the way down to 85%. Don't react to an emergency that doesn't exist.

# Alcohol

The effect of alcohol multiplies at high altitudes—two drinks on the ground equivalent to three or four at high altitudes. Even small amounts of alcohol could impair the helicopter pilot's judgment and decision-making abilities. As little as one ounce of liquor, one bottle of beer, or four ounces of wine could impair flying skills, with the alcohol consumed in these drinks being detectable in the breath and blood for at least 3 hours.

There is no way to increase body metabolism with alcohol or alleviate a hangover (including drinking black coffee). At night, you should scan your surroundings slowly to permit the off-center viewing of dimmed objects. Helicopter pilots need to look at their gauges and instruments, which is about 2 feet in front of them. Peripheral (off-center) vision the most effective action at night.

FAA regulations forbid anyone to act as a crew of a civil aircraft within 8 hours following the consumption of any alcoholic beverage.

Medical conditions and over-the-counter medications are seriously degrading pilots' performance. Many medications have primary effects that may impair judgment, memory, alertness, coordination, vision, and the ability to make calculations. Others have side effects that may impair the same critical functions.

Any medication that treats depression, or the nervous system, such as a sedative, tranquilizer, or antihistamine, could make a pilot much more susceptible to hypoxia. You should not perform crew duties while using any medication that is conflicting with safety.

# Minimum fuel

A Helicopter pilot should advise ATC of his/her minimum fuel status when the fuel supply has reached a state where, upon reaching a destination, he/she cannot accept any undue delay. It indicates a possible future emergency but does not declare one.

The FAA expects the flight crew to monitor the aircraft's flight path.

Suppose the remaining usable fuel supply suggests the need for traffic priority to ensure a safe landing. In that case, you should declare an emergency due to low fuel and report fuel remaining in minutes.

# Collisions

An approach to a narrower-than-usual runway could create the illusion that the aircraft higher than it actually is. Wider-than-usual runways may result in higher than desired approaches. Leveling off too high not affected by the runway width, but rather by the Helicopter pilot's landing proficiency.

A near mid-air collision is defined as an incident and is associated with an aircraft's operation. The possibility of collision occurs because of the proximity of fewer than 500 feet to another aircraft, or a report received from a pilot or flight crew stating that a collision hazard existed between two or more aircraft.

An aircraft is in an emergency condition when the pilot becomes doubtful about his position, fuel endurance, weather, or any other condition that could affect flight safety.

Pilots of aircraft that produce strong wake vortices should make every attempt to fly on the established glide path, and as closely as possible to the approach course centerline or to the extended centerline of the runway of intended landing.

The potential hazards of a midair collision and the near midair collision emphasize the basic problem areas related to the human causal factors. Improvements in pilot education, operating procedures, and scanning techniques are needed to reduce midair conflicts.

# THE NTSB

The National Transportation Safety Board's function is to conducts investigations of all civil aviation accidents in the United States. The Safety Board has no regulatory or any enforcement powers. The Board is only analyzing the information and determines whether a probable cause exists to introduce it as evidence in a court of law.

Each investigator is a specialist responsible only for his part of the investigation.

The NTSB individual working groups stay as long as necessary at the accident scene from a few days to several weeks. Their work is then continuing at the Washington headquarters, forming the basis for their analysis, and drafting the report that goes to the Safety Board itself. It may take perhaps 12 to 18 months from the date of the accident. The Safety Board may issue safety recommendations at any time during an investigation.

Once The NTSB has complete discretion, which organizations it will designate to investigate. The investigation does not allow assigning it to individuals in the legal or litigation positions.

The Safety Board does not investigate any criminal activity but can refer the matter to the FBI. The FBI will lead a federal investigation, with the NTSB providing any requested support.

If the NTSB list flight controls malfunction or failure as an incident, it requires immediate notification to the field office. An aircraft involved in a collision is required to file a report only on request from the NTSB.

Under NTSB rule, Part 830, all information obtained from the flight data and the cockpit voice recorders connected with the investigation should help determine the cause of accidents or occurrences. The Administrator does not use flight data or cockpit voice recorders for any civil penalty or certificate action.

The operator, who has installed the approved flight recorders and the approved cockpit voice recorders, must keep the recorded information for a minimum of 60 days.

# Mid-air collision

Studies have shown specific warning patterns. Surprisingly, almost all collisions occur in daylight and under visual flight conditions. The majority are within five miles of an airport, in areas with heavy traffic, and on weekends where more pilots perform more flights.

The speed at which two planes have collided is relatively slow, generally much slower than both aircraft's speed. The majority of the air collisions result from an airplane flying faster and colliding with a slower plane.

It is impossible to say whether an inexperienced pilot or an older, experienced pilot is more likely to be involved in a collision in the air. A beginner pilot has so much to think about that he is bound to forget to look around. On the other hand, after spending many hours flying without noticing dangerous traffic, the older pilot can also become complacent and forget about scanning the sky. No pilot is invulnerable.

According to the NTSB, a severe injury does include severe tendon damage and second- or third-degree burns, covering more than five percent of the body. When an aircraft accident occurs, the pilot must notify the nearest NTSB field office by the most expeditious means available.

If you are looking for other traffic, you most often look straight with occasional glances left and right. This way, you protect yourself only against 5% of the most common collision scenarios during the flight.

The NTSB has expanded the definition of a collision to include runways.

45% of the collisions occur in the course of traffic. Two-thirds of these occur during the approach and landing when the aircraft is in the final approaches or over the runway, confusion about the plane's position, and their landing order often begins earlier.

Some of the most severe ground collisions occurred at airports without a control tower and in elevated airports with a confusing number of taxiways.

# FLIGHT PLANNING

As a pilot, you are responsible for all aspects of the flight and related information, i.e., catering, ground transportation, servicing requirements, reservations, etc. You may assign this duty to the other pilot but retains the responsibility for the task. There must be no confusion as to which pilot does this function. Both helicopter pilots will thoroughly review the trip manifest, prepared by the dispatcher. Any discrepancies or questions should be reviewed with the dispatcher. Confirm the number of passengers on each leg so that the proper fuel planning could be accomplished and ensure all ground transportation needs. Heliports are not normally included in the **NOTICES TO AIRMEN** reports issued by local flight service facilities. It is, therefore, your responsibility to determine the latest status of the destination heliports. The dispatcher will make every possible effort to determine the heliport condition before the flight takes place and relay appropriate information to the crew.

There will be some occasions like holidays, weekends, change of destinations, winter snows, dignitary or politician closures, delays, etc. when you will be required to ascertain the heliport conditions on your own. Heliport conditions should be confirmed at least 30 minutes before arrival with a call to the destination heliport or the ATC for the latest advisories.

If a phone call is not feasible, try to make radio contact as early as possible into the flight to allow a change of destination if conditions are warranted.

The landing authority at privately operated heliports will always be arranged through the dispatcher. A contact number to determine the heliport conditions will be provided to the pilot before the proposed flight departure.

Usually, the senior helicopter pilot will fly the first leg as PIC (Captain) when departing. Subsequent legs shall be alternated under operational qualifications and by mutual agreement between pilots.

Flight plans documents are filed by a flight dispatcher before the departure, which indicates the planned route or the flight path. The flight plan is formatted specifically in ICAO doc 4444. Flight plans are comprised of information such as departure and arrival points. They also include the estimated time En-route, the alternate airports in case of bad weather, and the type of flight (whether instrument flight rules IFR or visual flight rules VFR), the Helicopter pilot's information, the number of people on board, and information about the aircraft itself.

An IFR or a particular type of VFR flight plan must be filed. For IFR flights, flight plans are used by the Air Traffic Control to initiate tracking and routing services.

For VFR flights, their only purpose is to provide the needed information for search and rescue operations or for use by ATC when flying in a Special Flight Rules Area. Flights may VFR, IFR, DVFR, or a combination of those types composite.

The flight plan blocks are:

• Aircraft Identification the registration of the aircraft, which usually is the flight or tail number.

• Aircraft Type/Special Equipment: the type of aircraft and the type of equipment.

• True airspeed in knots: The planned cruise in true airspeed of the aircraft-in knots.

• Departure Point: Usually, it is the airport's identifier from which the aircraft is departing.

  • Departure Time: Proposed and actual times of departure.

  • Times - Universal Time Coordinated.

  • Cruising Altitude: Cruising altitude or flight level.

• Destination: Point of the intended landing.

• Remarks: Any information that you believe necessary to provide to ATC.

One common is remark 'SSNO,' which means the pilot is unable or unwilling to accept a SID or STAR on an IFR flight.

• Fuel on Board: The amount of fuel onboard the aircraft, in hours and minutes of flight time.

• Alternate Airports: Airports of intended landing as an alternate of the destination airport. It may require for an IFR flight plan if poor weather forecast at the planned destination.

• Helicopter pilot's Information: Contact information about you for search and rescue purposes.

# VFR Flight Plan Instruction

➢

➢ Type of flight plan. Flights may VFR, IFR, DVFR, or a combination of types termed composite.

➢ Aircraft Identification: The registration number of the aircraft, it's usually the flight or tail number.

➢ Aircraft Type/Special Equipment: The type of aircraft and its equipment. For example, a Mitsubishi Mu-2 equipped with an altitude reporting transponder and GPS should use MU2/G. Equipment codes found in the FAA Airman's Information Manual.

➢ True airspeed: it's the planned cruise true airspeed of the aircraft measured in knots.

➢ Departure Point: Usually, the identifier of the airport from which the aircraft is departing.

➢ Departure Time: Proposed and actual times of departure. Times Universal Time Coordinated.

➢ Cruising Altitude: The cruising altitude or the flight level.

➢ Route: Proposed route of flight. The route consists of airways, intersections, Navaids, or possibly direct.

➢ Destination: Point of intended landing. Typically signify the identifier name of the destination airport.

➢ Estimated Time En-route: the time-lapse between departure and arrival to the destination.

➢ Remarks: Any information you believe necessary to provide the ATC. One common remark 'SSNO,' which means the pilot unable or unwilling to accept a SID or STAR on an IFR flight. Fuel on Board: The am.

➢ The amount of fuel onboard the aircraft has given as hours and minutes of flight time.

➢ Alternate Airports: an alternate Airports of intended landing.

➢ It may require for an IFR flight plan if poor weather forecast at the planned destination.

➢ Helicopter pilots Information: Contact information on you for search and rescue purposes.

➢ Number on-board: Total number of people on board.

➢ Color of Aircraft: The color helps identify the aircraft to search and rescue personnel.

# IMPORTANT NOTES

The Equilibrium in-flight exists when the sum of all forces and all moments around the center of gravity equal to zero. A state of equilibrium the absence of acceleration and could either linear or angular.

Trimmed flight exists when the sum of the moments around the center of gravity equals zero; hence, if you are in Equilibrium flight, you're in trimmed flight, but the reverse may not be valid.

Static pressure the force that each particle exerts on those around it.

Atmospheric static pressure is decreasing with altitude's increase.

Air density is the total mass of air particles per unit of volume.

Air density will decrease with an increase in altitude.

Temperature is the average kinetic energy of air particles.

At 36,000 feet, MSL, the Temperature constant at -56.5°C (Isothermal layer).

Lapse rate: Temperature will decrease when altitude increase, at a rate of 2°C per every 1000 feet, up to 36,000 feet MSL.

Humidity is the amount of water vapor in the air.

As humidity increases, the air density will decrease.

Air viscosity is the air resistance to flow and shearing.

The air's main gases are 78% Nitrogen, 21% Oxygen, and 1% other gases.

An increase in density altitude decreases aircraft performance. As the density Altitude increase, the power available will decrease, and the increase in power is necessary.

As Temperature increases, the speed of sound increases.

As Temperature increases, the density Alt increases.

Aerodynamic Center is a point along the chord line of helicopters, in which all changes in the lift force are considered to take effect.

Airfoil is a structured wing, designed to produce lift as it moves through the air.

The Angle of Attack is the angle at which the airfoil or fuselage, such as a helicopter rotor, meets airflow.

Chord Line is a straight line intersecting the leading and trailing edges of an airfoil.

Induced Velocity is the speed and direction of the induced downward flow of air, which results from an airfoil's passage.

In-Plane Drag force is the total of all the decelerating forces in the plane of rotation.

Lift is the component of the total aerodynamic forces (thrust), vertical to the relative wind.

Rotational Velocity is a component of the relative wind produced by the rotation of the rotor blades.

The rotor disc is the circle acclaimed by the blade tips in the tip path plane.

Rotor thrust is the vector sum of forces produced by the rotor system and is used to overcome the helicopter's weight.

Translational velocity is the wind correction made to rotational velocity once the helicopter is in motion.

Linear velocity is the combination of the rotational and translational velocities.

Symmetrical Airfoils is the distance from the chord line to the upper or lower surface of the airfoil.

Feathering is described as the blade's rotation about its span axis, and it permits changes in the blade pitch-angle.

Collective is the change of all blades in the same pitch.

Cyclic is changing the pitch on the blades, according to the azimuth position, with the opposite blade changing its pitch equally, but in the opposite direction

Flapping is the upward and downward rotation of a rotor blade about the horizontal hinge.

Centrifugal force is the outward force that is created by the rotation of the main rotor head.

When the rotor disc has tilted, the center of gravity of the rotor blades changed.

When the rotor disc has tilted, the down blade's radius increases, and therefore the rotational velocity for that blade decreases.

When the radius for the up-blade is decreased, the rotational velocity increases.

The main rotor torque will cause the fuselage to try and rotate clockwise.

The tail rotor uses 5-15% of the total power available.

The main rotor uses 85-95% of the total power available.

# Abbreviations

NAT  -North Atlantic

IFR - Instrument flight rules

VFR-Visual flight rules

ETOP - Extended range Twin Operations

NOTAM - Notice to Airmen

FDC NOTAM - Refer to information that normally is regulated.

ATC - Air traffic control

RFFS - Rescue and Fire Fighting Service

ICAO  International Civil Aviation Organization

ARINC - Aeronautical Radio, Integrated

ATP  - Airline Transport Pilot.

ILS - instrument landing system.

MDA - Minimum Descent Altitude.

DA - Decision Altitude.

DH - Decision Height

OTS - out of service.

WAAS - Wide Area Augmentation System

RNAV - Area navigation.

ADIZ - Air Défense Identification Zone.

IAP - Instrument approach procedure.

GPS - The Global Positioning System.

AFSS - Automated flight service station.

CAT - Clear air turbulence.

SIGMET  - Significant Meteorological Information.

AIRMET - Airmen's Meteorological Information

TAF - Terminal Aerodrome Forecast.

SQ-A  squall.

FPM - Feet Per Minute.

FPD  Freezing Point Depression.

GTG - Graphical Turbulence Guidance.

TIBS  Telephone Information Briefing Service.

HIWAS - Hazardous Inflight Weather Advisory Service.

AWOS - Automated Weather Observing System.

LLWAS -  Low-level Wind Shear Alert System.

DURD - during descent or on approach.

DURC - during climb or after takeoff.

KTAS  'Knots True Airspeed.'

L/D ratio-means the Lift-to-drag force ratio.

EDCT - Expect Departure Clearance Time.

STAR - standard terminal arrival.

AGL  Above Ground Level

IAS Indicated Airspeed

IGE in Ground Effect

INOP Inoperative

ISA International Standard Atmosphere

KCAS   Knots Calibrated Airspeed

KHz   Kilohertz

KIAS Knots Indicated Airspeed

KTAS Knots True Airspeed

OGE Out of Ground Effect

UHF Ultra-High Frequency

VHF Very High Frequency

# The Metric System

1 centimeter = 10 millimeters = .39 inch.

1 meter = 10 decimeters = 39.37 inches.

1 dekameter = 10 meter = 32.8 feet.

1 kilometer = 10 hectometers = 3,280.8 feet.

1 gram = 10 decigram = .035 ounce.

1 kilogram = 2.2 pounds.

1 liter = 10 deciliters = 33.81 fl. Ounces.

1 dekaliter = 10 liters = 2.64 gallons.

1 hectoliter = 10 dekaliters = 26.42 gallons.

# THE END

www.ingramcontent.com/pod-product-compliance
Lightning Source LLC
Chambersburg PA
CBHW021350210526
45463CB00001B/56